9/11 Attacks...

First African American President...

Trump Foretold In The Secret Hitler Files?

A Concerned Citizen

Preface

I was on my way to New York City by train to visit a female friend on the day of September 11, 2001. The train was stopped at Trenton, New Jersey because there had been an alert that something would hit the World Trade Center. I still remember my friend's warning: "Don't arrive downtown too early because today is going to be somewhat different from a usual day."

The best thing to remember is not to carry hatred for anyone. Because everyone has had tragedy in their life.

But I wonder, were there any psychic predictions of the events of 9/11 that were made years ago? And can anybody's fate be changed by a psychic prediction? Also, are there actual gods who can move people? And what are metaphysical people?

Contents

Chapter 1

Hydra's Manifestations Of Demons Within The United States, Europe, And All Other Places

{9/11 Attacks from Occult Activities or Great Britain's 5/11 Curse? "Don't Use Those Aircrafts" in 1948?" What are Hydra's 20 Medusas and the 2300/44 Curse?... Warnings from the Florentine and Dresden Codexes? Mayan king, in 611c.e., forecast the catastrophes prior to 2012-2021a.d.? Ancient Mayan numerology and the 21st Century: .131313 divided by 11 or .01193754545?}

There is a forty-two-year-old man named Dennis Martino who is concerned about the public worldwide. Some people say that his dark and wavy black hair gives his Middle Eastern and French heritage away. He lives with his girlfriend, Gloria, in Greenwich, New York. Dennis was named after his father, who is also his ex-boss, having worked under him at Greenwich Community College. Dennis is a teacher of world history and ethnic, cultural studies, like his father. Greenwich Community College voted him Best Teacher of the Year for his coverage of the 2008 presidential election in his classes. As he accepted the award, he gave respect to his ailing father.

Dennis had grown up in the Greenwich area, so he has always been remarkably familiar with and concerned about the townspeople there. He has worked and campaigned for Donald Trump in 2016, and prior to that, he had been part of the 2008 John Edwards presidential campaign, working hard campaigning in Greenwich and the area around the Adirondacks in New York. He's been slightly disappointed when scandal overcomes, resulting in John Edwards losing the race.

Dennis didn't get involved in the 2012 election, but he did get election fever in 2015. So, when the school season approaches—classes are scheduled to begin on September 11, 2015—he begins to anticipate covering another election. He supports Donald Trump's candidacy. When he begins to express his feelings to his colleagues at the school, a few previously unanswered questions about the 2008 presidential race emerge. He has assembled some of the same teachers he worked with back in '08 and recruited a few new people to join his campaigning team.

One day in the spring, as the group is comparing strategies in the teachers' lounge, a new teacher from LA begins to speak, "My name is Harvey, and I'm one of the new European and Mesoamerican history teachers to this school. I've lived and worked in Los Angeles, California, for the past ten years. I must admit that prominent Los Angeles families who were involved in the 1972-73 Watergate scandal and the 9/5/75 attempted assassination trials of President Gerald Ford have always whispered the name 'Obama'! I still analyze statements from Geoff Shepard and others about the government's future predictions.

Thirty-two years after 1945, on September 13, 1977, the Richard Pryor show aired a comedic episode about a 'First Black President', or the 40th leader of the United States. During this time Hollywood reported that he was placed into that position to 'Save us Demons', in America! And comedian Paul Mooney had always whispered to his audiences back in 1982, about President's Ford's comments on the Philadelphus experiment in the 1940's! I clearly recall Paul Mooney's predictions and beliefs about Obama and Hilary Clinton for 2008, on the 3/27/07 David Letterman show! Back in Los Angeles people around Mr. Mooney's comedy café have always whispered about a Black president's arrival in the Twenty-First Century! Since the 1977 Richard Pryor episode, and Mr. Pryor's appearance on the 1979 Tonight Show people have discussed about how government officials psychically maneuvered 'The Little Brown O.B. A., or Obama Boy' out to Los Angeles, C.A.! And did Sharon Tate and others try to understand the 'Little Boy with Ptolemy's Eye, President of the Unseen World, and demons in 1967?

But immediately after Mr. Mooney aired on Letterman's 2007 show, famous radio D.J.'s from L.A., N.Y., and D.C. discussed about how the 'Influential and Wealthy People' used Magic, or Kali's deities to foretell the future! People in Los Angeles didn't understand that in 1918 people from Tsar Nicholas Romanov's family forewarned the public not to use any paranormal or 'Out-of-Body' psychics! Nevertheless, the biggest questions psychic followers had in the seventies and well into the nineties were: [1] What was Montezuma II's Revenge? Were Demons from the past trying to Kill Everyone? Were Obama and others sent to investigate the Genies, the People with Vision, or the Watchers from the Book of Enoch! This sounds like a real fantasy but my parents remembered Vice President Spiro Agnew of the United States conversations about a weird 'U.S. Government Scuffle' around 2019 -2021! Diplomats like him were told to squeal on these Ptolemy's futuristic eye predictions! Some Los Angeles residents didn't believe these myths teased Mr. Agnew and wanted this great disaster to occur! Former President Reagan born in 1911 was the one prophesied to explain the 9/11 curse. When Reagan became governor of California in 1967 -1975, he threatened to Get Rid of these political demons, and liars.

Quite a few Hollywood insiders suggested that some of these episodes derived from the psychic's Philadelphus Experiments during the 1900's- 50's. Some of the psychics claimed that these predictions would occur within 24 and 42 years after 1977, or the beginning of the Soviet-Afghan War in 1979-1989! This is where I started doing my research in the occult fascination of the 55th Legion within the years of 2001- 2019. Maybe, someone would like to explain these weird dates: 1945-1977= 32, or {3+2=5}. When you add the 24 and 42 years together from 1977 you'll arrive at these results '2+4= 6, and 4+2=6, or 5-6-6! Theres's proof that these European and North American demons knew of this weird dates and numbers!

But before I continue everyone should reinvestigate the 1961 U.S. presidential inauguration and L.B.J.'s weird occult statements about the 1967 Israeli 6- Day War. They're historical testimonies from King Hussein of Jordan which stated: "These Sighted-Witches put a Demonic-Hex sign on them! In Jewish Kabbalah the demonic 11th., +15th., +18th., Titans will line up and form a deadly trio and destroy nations in the 44th year or 45th president!" Just look at Tecumseh and William Harrison in 1841, Franklin Roosevelt in 1944, and possibly the U.S. 45th president! Why did the Israeli 6-Day War end at 6 PM. on June 10th? Why all those 6's? Most of the demolition and destruction occurred within June 9th-11th, in 1967. Some called it the small apocalypse of the 6's! King Hussein has always stated who was being picked- on within the next 40 years? Was this Lord Shiva from the demoness Kali?

I'm glad I still remember passages from the Epic of Gilgamesh and the ancient Book of numbers: 1666-1999, 1499-185?, and the strange Great London Fire on September 6? Was the London Fire preplanned with all those 6's? Was this apart of Montezuma's curse 11/8-9/1519? Who did Aztec priest summon in 1520-2015? Did the Ancient Aztec predict the 'President of the Unseen World' approximately 6 years before 2015?

Epic of Gilgamesh's protection from Lilith and Kali's demons

In college I studied kabbalah, Mesopotamian history, and the mystical demoness Lilith! Some scholars of mystics believe in the past and future predictions within the epic of Gilgamesh written between 2100-1200 BCE! Historians still debate whether this 'Deluge' already happened or would occur in 2100 or 2001ad.? When I examined the Burney Relief I personally think its some type of idol of Lilith a feared Night Demon that used to ravage Samarian, Babylonian royalty, and other families! There's also Kali's demons designated with the forbidden flower.

Many of the psychics believed they were getting their predictions from Tecumseh or the Aztec priest in the 1500's. Many of W.W.2's Navajo code breakers still remember the Shawnee's or Tecumseh legends of the Blue Jackets and Red Sticks. This sounds weird but they believed these Aztec priests used demonic forces they called 'Hydra's Demons', or the 9th and 11th Gods! Ronald Reagan and others believed in this 1840 Tecumseh curse, and the 2012-21 Aztec curse on the government. Government officials throughout 1977- 2015 believed in a weird government coup de tat in 2021! Could they have predicted the militia's revolt on 1/6/2021 caused by Trumps defeat? Was this the 'Government Scuffle' that Mr. Spiro Agnew talked about in 1977? I know President Trump warned me to write about some of these predictions in 1988!

Can you believe former President Ronald Reagan's attempted assassination by John Hinckley Jr. on 3/30/81 was predicted by Ptolemy's eye in 1975 or earlier? Unbelievably, since the late 1800's they've heard tales of Hydra's demons, and spirits being used around Ardmore, Oklahoma, and other Native Americans reservations in the area. Fortunately, former President Reagan devoted himself to these predictions. I strongly recollect Mr. Reagan, Bob Hope, and Emmanuel Lewis's public speeches about gun and crime control in 1985. This group warned the country about a viral outbreak in 2019 and told the country not to spread viruses to others! In my circle of friends, we've always talked about the year 2001, or twenty-four years after 1977. And government psychics who worked in the secretive Philadelphus, and New York Experiments predicted a young ethnic man would become a Chicago senator 24 years after 1979, or 2003! I still review some of Carter conferences on the beginning of the Soviet-Afghan war. Did psychics in 1961 correctly foresee the 44th United States President's future forty-four years later, in 2005?

But let me begin by explaining the connection between politics and the fascination with the occult. There has always been a big cover-up of the '60's actress Sharon Tate's premature death in 1969! Most investigators believe that her 1966 movie 'Eye of the Devil' enticed a large occult crowd and fueled their belief in Hydra the 9th god, and the 13th angel. When they filmed in Britain and France, they've always talked about their 5/11 curse, the Aztec/ Mayan curse of 2012-21, and the ancient Mayan numerology code: {.13 divided by 66= 00196969696}. Was there a connection between the current British royal family and Ms. Tate's movie? I still have two unanswered questions: [1]. Why was that movie released 13 days after 5/11, or Guy Fawkes Day, and [2]? Was Princess Grace Kelly of Monaco's 1982 car accident predicted to occur 15 years before 1997? Weren't officials aware of the 1597 or the British 5/11 curse? People should be aware of the deadly 9th god, Hydra's demons. Hydra's psychic heads extend themselves far are deadly to royal princesses.

You must excuse me, but people have said that there has always been a strong political message behind this movie. I'll give you a quick synopsis of this deadly 2012 Aztec Curse on American politics. Experts believe that a protective American president would arrive approximately five years before and after 2012! They didn't want this small lethal apocalypse to occur eleven years before 2012. We must investigate this weird cursed '6+2012+6' especially, after W.W.2 sixty- five years before!

Since the Moorish King Boabdil surrendered his lands at Granada to Spanish forces on Jan. 2, 1492, there's been rumors of a 'Moorish Othello' and demons, in the twenty-first century. These tragedies in the 1500's would begin with the arrival of Hydra and her 6 demonic deities, or [9+6=15]. This 'President of the Unseen World' was to battle the deadly 9 headed Hydra with her army of 50-60 demons! This 9 headed demonic snake would come out of Babylon and meet the eleven demons in England and provoke Britain's 5/11 curse! These are the same eleven demons who'd set up the Aztec government and create the disastrous meeting between Cortes and Montezuma on 11/8/1519! So, as Columbus and others searched for gold in the Moors hidden Mexican and Songhai government's rumors of a 9/11 in the New World began!

Some Kabbalah investigators believe that a 9/11 tragedy occurs when the Eleventh gods from the East meet the Ninth demons within America. The city of New York was chosen because it was considered the center of the original 13 demonic colonies. Surprisingly, when you study Mesoamerican numerology, you began to decipher the Mayan disaster prophecy of the: {.131313 series of numbers}. These series of thirteens should foretell Europe's, Thirteen colonies, South, and Central America's future in the twenty and Twenty-First century! The .13 series is used as: {.131313 divided by 11= .01193754545}. If you were to read the product backward, you'd receive these dates of historical events: {'54, / '45, /1937... 9/11, in 2001}! Students of ancient numerology use mirrors to study these ancient ceremonies. Remember the Aztec Curse of 2012 +'21= 33, and those digits are the beginnings of a small apocalypse! Ancient Mesoamerican numerologists used the 33rd legion and the fraction 9/11 within their demonic rituals to eliminate their enemies. There are two 33 legions, and they must unite every 66 years and create havoc!

Remember in Kabballah, the 9/11 fraction and the 18 devil's terminology are numbers used to divide a country or people, for example: [9 divided by 11 = .81818181818, or 18+6=24}! Those black magic numbers are similar to a curse of Ham in Genesis, 9:19 – 9:25, or the dangerous 'war and viral curse' of Exodus 9:11 – 9:14. The Exodus 9:11, 'war and viral curse' is usually used to chase away the evil magicians and their genies from the public. Sometimes the public sicken with a virus within eighteen years after this ritual. The ancient cross symbol of death is initiated in these 9/11 ceremonies... Was the 9/11 tragedy arranged to occur eleven days before the first day of Fall 2001, or 9/22/01?

To most numerologists, each alphabetical letter has a numerical number in the alphabet; therefore, 11/9 spell: {11=k, 9=I, 12=L, 12=L, or Kill over 24}. The equation: [L+L=24] are equivalent to the two letters Ls for the last pylon or world trade Center bldgs.

Nevertheless, most people forgot about Ms. Peg Entwistle's 1932 group of concerned citizens and the 1940's top-secret 'Philadelphia Experiment of 2012 – '15, and 2021'! Ms. Peg Entwistle was a British female actress cast in the movie the Thirteen Women, circa. 1932. Some of the cast members of that movie and others had strong occult beliefs in Native American black magic and historical predictions.

My grandmother was a Los Angeles nightclub dancer in the late 1920s and early '30s. She's the one who'd got me interested in Kabbalah and Mesoamerican belief in astrology, numerology, and the 20 Medusas! The United States government used to use psychics to channel the dangerous Mayan gods such as {Baktun the 13th, Ahau the 18th, and Chuen the 15th} to predict the future!

Many Native Americans and others informed the government and Peg's group about these approaching disasters described in chapters 4 and 5 of volume 12 in the Florentine Codex. Therefore, the Dresden Codex was hastily given to Dresden and Berlin Germany! Within the first twenty- two and three pages are predictions and curses for the current time period up until the Twenty- Seventh century. You must really understand the 51 or 52 almanacs.

Some Nahuatl priests label those chapters the dangerous Nine sacrificial chapters or 9/11 of the 2013 Aztec curse! The Florentine Codex is a combination of ancient Mayan/Aztec writings and is something like a Holy Grail to many Mesoamericans. These writings and myths were collected and given to Father Bernardino de Sahagun between 1540-1585 in Mexico City. Father Bernardino tried to rewrite these Aztecs' books and hieroglyphics about their beliefs and warnings. Father Sahagun was called by Hernan Cortes and others because he believed in the 1918 Spanish flu and the disaster eleven years before the 2012 Aztec curse! Something bad, very demonic, was going to occur 66 years before

2012! Was Hitler this Quetzalcoatl god and his demons born in 947 or 1947- 1974? And why are the Quetzalcoatl god's demonic twin brother Xolotl so similar to Romulus, Remus, and Venus? Were these demons supposed to arrive back in Mexico or Mesoamerica within 1519 -1915? Cortes may have predicted the 2019 viral pandemic! There's been many Mestizo legends and predictions about a futuristic disaster after the 1519 smallpox outbreak!

But Cortes was tipped off by his Indian guides that the 20 medusas and their demons would try to kill him and drive him out of Mesoamerica, so he went back to Spain and caused war upon the Aztecs on 8/13/1521. Maybe, he knew about the legend of a Mexican 9/16 or Mexican Independence Day. Many Spaniards like Cortes and Diego De Landa tried convincing the Native Americans that they have an "out of body" phenomenon and practicing witchcraft is not natural!

Many Native Americans still put 'Omens' on people and turn humans into 'Nemos' or sacrificial objects. But the Nahuatl priests predicted the arrival of an Italian and Greek family who'd bring the 'Prince of Peace' along with a Black 'President of the Unseen World' in 2012, or four years earlier!

Remarkably, the Mayan Dresden Codex predicts calamities supposed to occur 66 years before 2011. And there's another slight sacrificial tragedy described in Florentine's chapters 4 and 5, eleven years before 2012. Was Osama bin Laden's assassination date of May 2, 2011, or 5/11 supposed to end the Aztec/Mayan 2012 curse? These ancient warnings came from Mayan priests under the rule of Pakal the Great, in 611c.e., in Palenque, Mexico! I'm still trying to understand how the ancient Mayans knew about Venus, the Roman goddess, Greek Delta Oracles, Memphis, or Philadelphus?

But even today, many Spanish and Mestizo people still believe in these ancient curses on intruders who steal Mayan/Aztec gold and land even after the Mexican religious inquisition! Most readers of the Florentine Codex believe they are moved by the Maya's gods like Baktun 13th's demons! But these Mesoamerican demons are the same demons that destroyed much of the Maya in the 9th century!

But by the late 19th century, presidential families started to produce movies about horrific predictions or presidential demons of the future, such as Ms. Entwistle's 'Thirteen Women movie. But before these apocalyptic threats of Hitler and Bin Laden leaked to the public, her body was found at the bottom of the Hollywood sign on September 18,1932, or 9/18. Did these strange European predictions or 'Mayan curses of .131313' lead to Ms. Entwistle's premature demise approximately four years before Adolf Hitler comes to power? Did Ms. Entwistle's New York psychic forewarn people not to live in New York in 1932? Were these the beginnings of the Nostradamus predictions? Was Franklin D. Roosevelt warned about the deadly poliovirus during the presidential election of November 6, 1932?

But these questions about the '08 presidential election aren't new. In 1974 and earlier, our government knew about an American 5/11 attack and a Black man becoming president! In the late 1980's I used to work part-time and vacation in New York City. There were always John Belushi's 1976 predictions of a 2019 viral plague and rumors of Billionaire Donald Trump becoming president within thirty years. He was predicted to preside over the country in the Chinese zodiac year of the Rooster and to lead the evil spirits away from the United States government. But his term would be made into a collusion spectacle by being elected in the zodiac year of the Monkey, just as President Clinton in 1992. On October 7, 1999, most New York pubs held open conversations on Larry King's cross-examination of Mr. Trump about his future vice-presidential candidates. Did Mr. Larry King suggest, "Either Oprah Winfrey or the prophecy of some Black man?"

And I've personally talked to many New York police officers about the weird predictions surrounding the late John F. Kennedy Jr., 7/1999. Who was Mr. Kennedy Jr. giving personal little personal messages to while the camera was filming? And did Saturday Night Live's comedian, John Belushi, warn the public about a dangerous car accident for a popular New York Spanish female radio DJ on 11/4/2019? Did the dangerous 33rd legion contribute to Mr. Belushi's premature departure at age 33? There's a lot more to that tale than they're telling the public. But I'm still analyzing personal testimonies from guests of the Jimmy Kimmel show, which aired on February 23, 2018, about Oprah Winfrey's political prophecy?

"In 1987, I personally watched average New York citizens being paraded around the offices of the Ed Sullivan Theater in midtown Manhattan. Everyone openly discussed The David Letterman show's predictions within the next thirty years.

"December of 1988, a friend from New York City invited me to read newly discovered letters from Sir Isaac Newton from 1704 and earlier. Apparently, Mr. Newton and his colleagues left documents behind which dealt with the occult legend of Othello, or the Omega genies, and witchcraft! I believe the Florentine Codex of 1577 had some influence on sixteenth-century religious texts and manuscripts. But some of these documents contained astounding information on the impending danger of world disasters in the 20th and 21st centuries! Back in the 18th century, many English citizens like Mr. Newton used to decode the Ptolemy's and Berenices' family biblical Revelations for the 21st century. Most of these Biblical texts and letters are written in Latin and are studied today while they remain in the Vatican City libraries and Egypt.

Many Italian Roman Catholics believe that the Nostradamus predictions came from the Observations of Daniel pg. 123, or 'six.' You'll comprehend the 21st century if you read Newton's Daniel Observations of page 'six' and his Occult readings for 2034 and 2060. There're many Italian and British scientists who proclaim 'Witchcraft' was being used within the last four hundred years or earlier! If you'd subtract fifteen and eighteen years from 2034, you'll have many 2016 - 2019 predictions! I still recall Denise Matthews, or Vanity's instructions for a young 'Ptolemy's Eye' to help explain the 33rd. Legion's predictions and warnings!

"Some of the documents written in 1704 clearly share information about 2004! Some of these documents appear to be translations of Ptolemy's warnings of a dangerous Succubus and its daemons. This Succubus and the Thirteenth Angel will cause World War II and the modern war in Iraq! Most of these papers are written in Latin, Greek, or Aramaic. Possibly, King James VI of Britain, Nostradamus, and others were trying to translate the Biblical teachings of Tower of Babel, Genesis 11:1-9, and Daniel Chapter 2. Was the 9/11 World Trade Center's tragedy in the United States prophesied in Genesis 11:1-9, or 11:9? Is the United States considered the New Babylon or Hattusa in the Bible? Are Hydra's Antichrist genies in America?

There's a Greek and Egyptian book on witchcraft and demons that predates the Bible called the Book of Nod. Some occult members write Nod backward, and that spells to "Dan." In the book of Nod, these two incidents have always been linked together and titled: Rebuild in Bezaliel's Thirteen Colonies, Chapter 9 verse

11, or 9/11. I've read passages where Thirteenth Angel or {The Shadow of God} sends the Ninth Angel, Bardiel, to rebuild after a fallen metal statue or towers. To some Kabbalah followers, the small '00 or '01 yearly time measurements mean that 9/11 is the first part of a series of attacks!

"After the first attack on the World Trade Center's on 2/26/93, I instantly recalled the British 5/11 sighted-dybbuk curse! In Great Britain, during Guy Fawkes Day, there's usually a conversation about the Moorish folklore of 1492, 'General Othello' predictions for the 20th and 21st centuries. Even Sir Isaac Newton shared similar beliefs with his occult predictions of 2034-2060! The General Othello legend of 1605 or the Omega jinn predictions came from earlier Middle Eastern writings which stated an indigenous African would rescue the country from these dangerous sighted witches in 2005! These dybbuk witches would bring the 1665 plague and attacks on government buildings. You can still read these future warnings and predictions about plagues and wars from King James VI of Scotland, the sighted witch hunter! One sighted-dybbuk witch claimed she was one of the eighteen devils who'd dwelt in England. For her religious persecution, the sighted-dybbuk witch threatened the British government with an eighteen-month viral plague! The sighted-dybbuk witch was believed to foretell the 1665 Great Plague of London.

"Most 20th Century television shows in Britain and New England recorded government conversations, predictions, and threats about future tragedies. King James VI wasn't alone with his dislike of these sighted dybbuk witches. When I read the Nostradamus predictions and Sir Isaac Newton's occult readings of 2034-2060, I begin to realize a dangerous pattern. If you subtract eighteen years from 2034, I strategize a 2016 viral plague. Did 18th-century psychics predict the 'Sound of trumpets' with the "Death of the Purple one" in 2016? Death, viruses, and presidential corruption usually come after a war subsides in the 18th year. Will this be the sign of 18 devils!

"They're many conversations on predictions about the US presidential elections for the 21st century. The assassination of Osama bin Laden, the two World Trade Center buildings, and the city of Nineveh, Iraq, has always been discussed. It's believed the biblical passage of Genesis 11:1-9, or 11/9 and Daniel Chapter 2 in revelations. When I studied the 27th book in the Bible I began to accumulate the 9/11 predictions about the Antichrist genies in 21st century America. Some followers of Kabballah reduce the Book of Daniel to a single digit such as: {2+7= 9}, or 9/chapter II.

"President Wilson's administration warned the country about the dangerous 5/11 curse throughout the 20th and 21st centuries. There's a 2009 prediction of a 'red-coated specter' who attacks people after his death. Was this the same demon who'd attacked Pope Benedict XVI while in St. Peter's Basilica? This red-coated specter has always been associated with the dangerous seven-headed beastly women from Turkey. These Eurasian women will bring death along with a deadly eighteen-year Iraqi war in the 21st century. These female demons will bring pestilence to the public towards the end of the war!

"I still examine these New York comedians' taped conversations and jokes about President Ford and Governor Carter in 1976. Many of the jokes forewarn the public about getting vaccinated in 2019 for a viral plague! These predictions came straight out of the United States government and Hitler's Germany! Did government psychics in 1976 or earlier predict a viral plague within six or seven years after the assassination of Osama bin Laden, or {2011+7= 2018}? In 1917 the Russian government sent strange warnings that groups of people who call themselves demons or the 18 Devils were headed to Los Angeles or America. Are these the same demons the ancient British called their magical 'Druid' priest? The Romanov gave this country a list of predictions for the next one hundred years! The Russian psychics chronicled that 'Nothing would stop the 18 devil's viral pandemic of 2019 after the 5/11 war! Back in the seventies, people used to talk about the government coverup of Antichrist Genies or the nasty manifestation and its demons!

"When I worked in New York City, I've investigated audience members and average New York citizens about Phil Donahue's "Holocaust Debate" show in 1994. Some audience members reported that immediately after the show ended, seated protesters and others began shouting warnings to the guests and public of impending danger within six or seven years. Someone began comparing analogies between Great Britain's 5/11 plot and the February 26, 1993, attempted World Trade Center truck bomb! These two historical incidents are very similar and should be studied.

"Odd statements from Mr. Donahue, guests, and average citizens such as: "Don't use those aircraft on us! ... Please tell us more about 'Hitler's psychic ability,' mind control, and demons. Some of Phil Donahue's guests agreed that Germans followed Hitler because of abnormal abilities. 'He was Sighted and could see through the Camera!' It's astounding those British generals in 1948 were informed that aircraft would be used during the next US attack on public buildings!

"Whenever the public discussion of Hitler's Holocaust arose, Hollywood movies such as Village of the Damned, in 1960 and Children of the Damned, in 1964 were examined. These stories by John Wyndham and the European government in the 1950s had done extensive research in paranormal activity. They were trying to comprehend psychic ability in 'Sighted People.' Do premature death and political calamity occur in the Chinese zodiac years of the Snake and the Stalking Tiger?

Do ordinary people have genies or demons inside them? Did the German and the Russian government inform the British Government of an 'Ago of Nigeria' who'd become president within forty years? Or will a blonde female succubus who experiences out-of-body activity become a 21st-century psychic or Druid! Those demons who'd surround her are very dangerous. This story is like the Polish Jews eyewitness accounts of weird psychic seances in 1934. We must remember the strange demons of the Armenian holocaust of 1915 or {5+1= 6}. What kind of daemons or possessions surrounded the Eurasian Dybbuk box?

"Friends in Los Angeles have always said that the Lyndon B. Johnson administration had informed the public that a young man like 'Ago,' born in 1961, would become president within forty-four years. They were even documents on how they hid 'The young brown boy' in a distant American Province. My Russian friends have said that Ago is initials written backward. You must be able to read bidirectional to analyze many of these ancient texts. But if you use a mirror, you can clearly see the initials OBA. President Obama and the Ptolemy were to be called to the United Kingdom to help explain and comprehend Hydra's Antichrist Genies or Demons! People must remain observant of the giant triangular vortex on this country. This vortex spans from New York, Los Angeles, and the Hidden, City.

But I'm still trying to explain the connection between demons and viruses. Vladimir Lenin and the Bolsheviks used sacrificial peasants like Khioniya Guseva to help explain the 'Antichrist Genies' in their Russian revolution. Certain Russian monks such as Sergei Trufanov used to infect male or female prostitutes and peasants with syphilis and tell them to assassinate Russian leaders! Just read the Pokrovskoye, Siberian police report on the 1914 Rasputin stabbing.

"If you've read the Russian police report, Ms. Guseva apparently tried to explain strange cults, time, and Demonology. Words and phrases such as Anno Domini or 'AD' mean 'In the year of the Lord'. Some of Guseva's statements compared with Jewish Kabbalah members such as: Did '00 AD start with Egypt's twenty-fourth Dynasty and the collapse of Thebes, Egypt in 663bc. Which means we're currently living in the 25th century. Some occult practitioners claim the succubus of Babylon and its demons are in control of world governments. Because the numbers 2400/563 amounts to the Devil's Apocalypse!

"Are Rasputin's or Hitler's demons currently trying to destroy everything! The medusas and the succubus came from the mythical land of Nod, or Antioch,

Turkey. You can read about the Land of Nod in the Bible! But some believe the 2009 election is really written as 2409. Hitler's demons were supposed to come approximately sixty-six years before 2400, or 56/2400. Sixty-five years before 2400 is 2335.

"Remember in some Kabbalah cults, the year is added up to single digits and is used in sorcery {for example 2+3+3+5=13}. The number 13 is represented by Bezaliel or the Thirteenth Angel! Apparently, some Roman Catholics believe that the ancient sorcerer Bezaliel created the thirteen colonies through witchcraft. You can still read ancient documents from the 'Cult of Janus' in Rome, Italy, and the Vatican City. Maybe the United States president is trying to control the Antichrist genies!

"But many still ask, "What can this Ptolemy See?" Within Hollywood's movie with Sharon Tate in 1966, 'Eye of the Devil.' British Royalty in 1967 didn't want to see the death of another English Princess thirty years later. And the British, Germans, and Russians wanted to protect their families against these demons. Some Kabbalah members believe that the Star of David is really the 'Eye of Ra,' which is sealed to protect us against the Hydra's Antichrist Genies! Italians and Greeks were sworn to protect the 'Eye of Ra' within those families of the Ptolemies. But is there really a dangerous pentagram on the Middle East?

"President Kennedy's weird dreams of a succubus who'd threatened his death was reported within months of the release of these two movies. Along with the collapse of a famous tower on the six children. The government wanted to assassinate Osama bin Laden and a female Druid demon before the age of six! Researchers of the Kabbalists cult know that English Druids are the priests in the early Celtic religion. Those English Druids used the Germans Goths and witchery to destroy the Italians. Is there a young woman with a dangerous Druid, Drusus, or Bastet demon currently living in the United States?

"Certain Native American cultures still currently practice the ancient art of black magic and the use of genies during ceremonies. But the Succubus of Babylon and her genies need to be carefully studied. In the 1870's Chiricahua Apache, Lakota Sioux, and Cheyenne tribes have reported that black magic was used on General Custer! Even today, some undeniably believe in the Shawnee or Tecumseh Presidential Curse! This Curse was reported to have 'cracked India' in half to create a Pakistan. Why else could explain the thirty-one-day term of President William Henry Harrison? Some of the Chippewa and Shawnee Indians believe that he was trying to control the evil spirits with his birthdate of February 9, 1773. Some Native Americans still whispered that a dangerous 9/11 war/viral curse was placed on President Harrison, that would explain the deadly American Civil war 18 years later, which ended in 1865. And that Tecumseh's brother and others are currently brutalizing others in an attempt for vengeance in the 1900-2000s!

"I interviewed a man who'd talked to Eddie Murphy before he sat in the Johnny Carson audience on 1/1/82. Mr. Eddie Murphy made jokes about weird assassination attacks on former presidents. Were there predictions of a black male president in the early 2000s? The show aired only three months after the Reagan attempted assassination! Was something making people attack the US presidents? Remember, flight 93 in 2001 was aimed for Washington, DC.

"Hollywood filmmakers should investigate the folktales about the Tecumseh curse and the use of Babylonian succubus during the Comanche's peyote rituals. People say that a Medusa is a woman possessed by many spirits. A woman in this possessed state usually holds séances and gets involved with black magic to understand the voices in her head. Some believe that black magic is what a Shawnee prophet used on the US government.

"And the public should know about Mary Todd Lincoln's psychic encounters and the manifestations that ended slavery. The US government has been quiet about these threats since the Civil War, but the Roosevelt administration had tried to warn the entire country. The Kennedy and Johnson administrations kept these manifestations at the forefront of people's minds, but Nixon's people either hushed up about the topic of lost interest in it. Hollywood rekindled the manifestation rumors when the police started to investigate the suspicious cult murder cases in the late 1960s. Hollywood's coverage of the Helter-Skelter Files piqued the public's curiosity about cults and the supernatural. I believe that one of the films I have from 1974 talks about a suspicious famous football player trial that was supposed to occur in the 1990s or 2000s.

A. How are the Dybbuk Box and the Gatchina Letter related?

"But these topics talked about regarding the 2008 election are those of the Nostradamus prediction, including the evil manifestations in the dybbuk box. Dybbuks are dangerous demons that touch you. I don't want to get too far off the subject, but the years 2008 and 2009 have always related to these manifestations.

"These manifestations and their demons were said to have plagued the Russians, the Germans, and the people of other European nations. These are the very same genies or demons that'd threatened Emperor Paul I of Russia. The Russian government and 'Mad Paul' left a strange warning letter to his Romanov relatives before his assassination in March of 1801. This warning letter is still held somewhere in the Gatchina Palace in St. Petersburg, Russia.

"The warning clearly depicts the presence of the 1940's antichrist genies and an '01 Curse! The Gatchina Letter is believed to predict the assassination of the US president in 1901, 1918 flu pandemic, and the arrival of an ethnic president in the

United States. I researched the Gatchina letter and concluded that a Romanov relative and an ethnic diplomate would explain the 2400-year apocalypse. Paul, I informed his family to keep the letter hidden until 1901, and they were predicted to perish 17-18 years later. Psychics around the Romanov family had warned others of a deadly plague around 1918. Historians interested in the Armenian Genocide of 1915 should examine this letter about the weird demons of Nod.

"The Nostradamus prediction occurred on 9/11/01, and within 18 years, Obama became president. Will deadly plagues engulf America within 18 years after 9/11? Grigori Rasputin's documents should be studied because of their current-day virus predictions. I've even found a prediction of a young ethnic basketball player's death in 2020! Somebody said that his career should be based on his 6 ft.6 in. height and span of about twenty-four years, or 24/56. Back in 2002, this basketball player told me to write about this evil Mayan Curse of 2021.

"Winston Churchill started to write about these manifestations. Some Polish Jews in the World War II-era tried to contain some of them in a wine case, or a dybbuk box, in 1938. People of the Adirondack region in the late 1700s named some of the devils Minerva (the witch) and Dodona. Some people believe that Mrs. Abraham Lincoln complained of these manifestations and their demons when she was in the mental hospital. A Baltimore psychiatrist recounted that Mrs. Lincoln told the Europeans to bring over a young man named Sammy. They were to use his eye to see those demons that would touch her and other people. Her call for this young man and his genies was documented by that psychiatrist and by Ulysses S. Grant when she was committed to a mental institution.

"This young man's grandfather was brought over to the United States from Alexandria, Egypt, or Greece. The US government had said that his genies would arrive shortly before the third manifestation of 2001. Reportedly, the young Greek man was an advertiser for the 1918 movie: 'Eyes of the Mummy Ma,' with Emil Jannings. Seven years after that, he had a son with a mulatto showbiz dancer in Carlisle, Pennsylvania area. He and his fellow actors were supposed to alert the public to the great 5/11 disaster that was to occur in New York City in 2001. Some of the Civil War generals and European royalty were told to leave behind personal writings or films for the children of Queen Victoria. Was there a strange black magic number eight placed on Victoria's family? These predictions and threats probably contributed to the Nostradamus prediction, Obama, and President Trump. I have some of these texts and movies on DVD back at my old house in Los Angeles.

"There are many written texts that'd announced weird stories about demons on slave plantations. Most of the data had come from eyewitnesses on Maryland and Virginia plantations. One demon was chronicled, in Onley, Virginia, as the holler and screamer whom people called Pan. Their historians had traced him back to the 1804 Haitian Revolution. Back in the 1940s, the US government believed that Pan had threatened to come back as 'the red-coated specter' in 2009. I don't know why this government doesn't tell anybody and tries to cover this up. A lot of the disasters and scandals we've experienced could've been avoided if people had known this information," Harvey says, glancing around the room.

As Harvey continued speaking, he's developed a bigger crowd of listeners. Before long, all seven teachers in the lounge are listening or contributing to Dennis's and Harvey's conversation about the 2008 election. Dennis, intrigued by Harvey's conversation but having many questions, quickly clears his throat and blurts out,

"I know we've just met, Harvey, but can I ask you about these predictions regarding the '08 election and these threats of a manifestation? I've always wanted the prediction explained. I'd like to see what the government foretold in the 1970s. And can you elaborate on the prediction of the red-coated specter that was supposed to appear in 2009? Are you speaking of the little fairy tale Peter Pan? I don't mean to take the conversation away from any election talk."

Harvey slowly gets up, walks over to the refrigerator, and grabs a bottle of water. Then he faces the men and says, "Well, yes. Some of these predictions had come from a tiny area in Banias, or Paneas, Israel. At the turn of the century, there were small cults of Pan whose priests were paid for their predictions of the 20th and 21st centuries. The city of Banias, Israel, has another slightly older name of Caesarea Philippi, Israel. Because of its historical predictions, it's nicknamed "The New Caesar Philippi, of 20th century!"

"But these predictions along with past rumors about the assassination of specter Osama bin Laden occurred in 1974! We must remember President Ford's second attempted assassination on 9/22/1975. I've studied the tapes from former President Ford and Sara Jane Moore. From having grown up in LA, I have heard and seen a lot of weirdness.

Why couldn't New York City avoid this great disaster or the Shawnee Curse of the 2000s? In addition, what were the Comanche's peyote rituals? My father knew President Roosevelt and President Kennedy, and he worked with Ronald Reagan. He always said that those presidents desperately tried to caution everybody about these manifestations by using the news media. However, some actors in LA tried to hush up the threat of the disaster and the manifestation. I believe the answer lies with small Hollywood groups who didn't believe the prediction or else seemed to want this disaster to happen because people back in the 1970s, just like people now, simply didn't care. The peoples in these small isolated foreign countries aren't responsible for these old curses.

"And why couldn't John Edwards or Donald Trump become president and find Osama bin Laden? Probably almost everyone in LA knew of the psychic experiments done in the 1940s and on into the 1990s. Most of the psychics' predictions of a black male president are still on tape in Hollywood's archive.

That's why I talk about psychics, out-of-body experiences, and politics because, in Hollywood, they're all linked. Sounds peculiar, but everything should be firmly investigated. Before getting to those questions, I'd like to add something quickly. In LA, there are many famous murder and rape trials still under investigation. Some suspicious cases may have involved actors or government people.

"Los Angeles's has many little secrets: Certain psychics or people known as sighted dybbuks people follow certain celebrities. This leads me to the thought that Hollywood is haunted. Some of the actors and actresses have different nicknames for themselves, like Apollo or Hermes. It has been reported that some of them have memories of past lives and will tell you that they see others like themselves.

"The FBI calls these sighted people 'metaphysical.' In addition, the government uses psychics to understand these supernatural entities. Nevertheless, do metaphysical people haunt LA, California? During World War II, President Roosevelt and the German government both said, 'We do not want these types of people around us.' Scholars have whispered that these supernatural entities contribute to the odd and dangerous behavior in secret Hollywood circles. I still can't understand this strong attraction to witchcraft among certain celebrities." Harvey becomes quiet as he sips his water and walks back to his seat. After a couple of minutes, he addressed Dennis's question: "Yes, I know, Dennis, that you want to see my DVD collection on secret Hollywood predictions and the dangerous, red-coated man who appeared in 2009 and who is believed to be Pan. Talk to me tomorrow. I'll have someone send me all my files."

Turning to the rest of the group, Harvey continues: "I spent a lot of years in LA and have heard a lot of scary stories and seen a lot of scary things. The folkloric sightings of Pan or Mercury are at the top of that list—the threat of these people killing specters or the red-coated specter that appeared in 2009. In some history books, they reported demons or dybbuks leaving Haiti after the 1791 revolt and going to Louisiana. Did these dybbuks leave Louisiana and move to Los Angeles?

"It's been rumored that they hop into the bodies of white people and move their mouths at will. People out in Los Angeles will let you hear from Pan in their psychic experiments. When I was a kid, people informed me that this type of specter was extremely dangerous. Specter Pan was to become a famous dancer and lead the public into the great disaster. In 2009 Pan would publicly demonstrate the

Curse when he left his body and pulled down a religious figure that was bigoted. But he is not alone, as he works with many other demons.

"I know this conversation sounds absurd, but everything I've told you about Pan comes from taped conversations with some entertainers of the '70s to the '80s," Harvey says, shifting in his chair.

Dennis listens to Harvey's warnings for about half an hour more before he clears his lunch table and goes back to class. He has three more history classes to teach before the end of his school day. He seems to breeze through his last class, as there aren't any tests. Once the last student leaves his classroom, he quickly gathers his papers for the next day, locks his classroom door, and walks out of the school and into the parking lot.

He lives about two miles from the college. Driving home, he thinks about what was said in the lunchroom. *Is this the proof everyone has been waiting for? Was there any funny business in the 2008 election that caused the Edwards scandal? And what is the connection between being out of body and a 2009 specter? Why shouldn't people come out of their bodies? And does coming out of your body mean you're possessed? What is a metaphysical person? Were 1840s plantation owners ordered to call their local police if anybody had out-of-body experiences?*

And what weirdness scared Ulysses S. Grant and the KKK? He knows he has a lot to tell Gloria and his mother, Sylvia.

As Dennis pulls into his driveway, he notices that there aren't any lights on in his house. His girlfriend's car isn't there either. Dennis doesn't have any children, but he and Gloria are planning on starting a family in June. He parks his car, grabs his briefcase, and heads for the garage. Once inside the house, he sets his briefcase down, turns on the lights, and turns on the television. While preparing himself something to eat, he begins formulating campaign strategies.

Within half an hour, he has finished eating and is now settling down in his favorite chair in front of the television. As he watches television, he goes over his students' papers and the plans for his 2016 presidential campaign effort. He wants to call Gloria and tell her about this LA teacher and the '08 election. Gloria works for a Greenwich-based magazine company called Bright and Early that brings foreign news to the United States. With such a demanding job, she usually comes home late and wakes up early. As Dennis reclines in his favorite chair in front of the television with a glass of water, he decides to give her a call. Within a couple of seconds, he dials her number.

Gloria immediately answers her cell phone, saying, "I knew you'd be calling, honey, since it's past four o'clock .... I'll be there shortly."

Dennis responds, "Yes, Gloria. I'm so glad that the school year will be over at the end of this week. That'll give me time to work on what's left of the 2016 campaign. Are you still going to take your vacation next week? I think my mother's plantation house in Olney, Virginia, is the perfect hideaway vacation area. As we help cultivate the tobacco, I can start some of the campaign operations around the area."

There is a slight pause as Gloria gathers her thoughts. Then she says, "Well, if you want to. Your mother and I could be the tour guides again. I could learn something while investigating your family's historical role in the Underground Railroad for the next two weeks. It's been about two years since we last vacationed in Onley. Last time, your mother showed us the houses of the Underground Railroad on the part of the route that went through Maryland and Virginia. She showed us other historical houses that, like your family's, were involved in hiding the captives before they took the route farther to Poughkeepsie, New York, or somewhere else."

Before Gloria makes another comment, Dennis straightforwardly says, "Aha! In any event, I was discussing the 2008 election with some of the other teachers who worked on the campaign with me. They are still looking for ways to improve their skills for the 2016 election. Today my friend introduced me to a new teacher from LA who has just started at the school. His name is Harvey, and he seems to have some intriguing observations of the 2008 election year. He says that there are groups of people in and around Los Angeles who had been waiting for the great disaster in the 'new city.' He also claims that these people had prior knowledge of Barack Obama, or some other black man, becoming president. Harvey says that a lot of the taped conversations about these matters came from psychic experiments performed throughout the 1970s. This knowledge has been known since the experiments first began in the 1940s or even earlier. Harvey says his information comes from tapes of a syndicated televised talk show aired in 1974. He says that the show's host states that only this black president would be able to find and assassinate specter Osama bin Laden. Also, Harvey has a list of taped movies and presidential conversations with similar information that he wants me to study.

"Can you make some calls and bring me proof of the election year of 2008 and the subsequent presidential inauguration of 2009? I know you have knowledgeable people around you. Maybe they can check out some of the 1970s comedy talk shows. This could help improve some of our skills for the 2016 election," Dennis says bluntly.

Before Dennis has time to make another comment, Gloria quickly responds, saying, "Dennis, I'm so glad you're bringing this up. Just the other day, Brian, one of my coworkers, interviewed people about the Nostradamus prediction and the

Shawnee Curse. However, some Comanche tribes used the Antichrist Genies in their peyote rituals! European occult followers call these nymphs the 18 Devils!

"He told me that he was making documentaries of Native American rituals with the occult, and that definitely included the Aztec Curse of 2012! Brian was focusing on the 9/11 terrorist attacks and Britain's 5/11 predictions. He said he and others in his department were in the process of writing an article about the origins of the Nostradamus prediction. Brian also said that he has been researching a lot of World War II footage and found that a lot of the information on current events came from prisoners of the concentration camps at Ravenbruk, Auschwitz, and Treblinka. The Jews warned us that the Medusas and her demons would come to the Americas and wreak havoc. Please, this is no joke. Try to find someone who can see and believe this. The psychics elaborated on some of the information.

"Yeah, I believe that Brian did mention something about a threatening manifestation. I'm not sure how it threatens the United States—or anyone.

"Know that I am on the job of looking into this topic. I'll ask Brian for more of his articles. He's supposed to share articles with me anyhow. I'll let you take a peek at them when I get home. I think it'd be best if you asked your father, too. He and his veteran friends have always shared stories about the Auschwitz and Treblinka concentration camps.

"Let me go now and clean up. I'll be home within the hour," Gloria says, hanging up the phone.

B. Romanov's and Grigori Rasputin Battle with Russia's 'Real Demons'? Tsar Paul I, of Russia Letter at Gatchina Palace, St. Petersburg. Was Czar Paul Petrovitch victim of the '01 Curse?

Having said goodbye, Dennis quickly hangs up the phone and then calls his father. His father is ninety-three years old, and his mother is ninety, so he's very careful not to disturb them too much. Growing up in Greenwich, Dennis traveled through the Adirondack Mountains with his father and his father's veteran friends. He's familiar with stories of the old specter involving the Revolutionary War and going through to World War I and World War II. But his mother might have memories of the 1860 tobacco plantation and Native American occult rituals? And were there any predictions of a "Prince of Peace"? These thoughts are on Dennis's mind as he waits for his father to answer the phone.

He is hoping that his father is still awake at this time, figuring that he should be because he always watches the eleven o'clock news.

Dennis Sr. sounds friendly when he answers: "Yes, Dennis? I just talked to your brother, Jerry, a short time ago.

"So, tell me, what's going on?" Dennis Sr. asks shrewdly.

"Well, Dad, I was thinking about helping with the campaign of 2016, and then some details of the '08 campaign came up. I need to know about some of your old war stories about the weirdness in the concentration camps."

His father responds by saying, "Well, as you know, I didn't fight in World War II, but some of my friends fought. Some of them engaged in hand-to-hand combat, and others helped to retain the captive German troops. Some helped the Soviet troops free the concentration camps. A few held high-ranking positions in the army, navy, or air force. When they've talked to me about the concentration camps, they've mentioned some weirdness in those camps. And mounds of letters about 'Real Demons' who'd had attacked Grigori Rasputin and the old Romanov family. Mr. Rasputin is believed to have channeled an ethnic politician in America in 2008! But Concentration camp documents were talked about during the Nuremberg trials of 1945.

"Working as a welder in the New York City, Baltimore, and Philadelphia naval yards during the war, I didn't see any fighting, but their stories were all similar. I've listened to many lectures by the German actor Mr. Theodor Emil Jannings. Mr. Jannings had many stories of demons and the last days of the Tsar Russian Empire.

Over in Europe, their captains had told them to kill any officer or infantryman with glaring red eyes, as these men were sighted dybbuks like Hitler. The Polish Jews in the camps yelled, 'Somebody is using witchcraft and trying to kill us! The Medici family in Florence and Tuscany should be called. Sounds like the predictions from the Florentine Codex. These are Russia's 33rd legion of Demons, a part of the dangerous 66th legion. And they're being led by the 20 Medusas... The prophecy states that these 33 legions of demons should gather in the United States within the 55th year of the 25th century. The entire 55-66th legion should be led by the deadly Spanish legion, and they will begin to erase everything! They put a whole lot of devils on us!' Then the German Jews yelled.

"Back in the 1940s, some European diplomats had put an alert out in New York, Philadelphia, and Baltimore. They had their local officer in charge put a call out for a young man named Sammy or his father, as they needed Sammy's eyes to see the demons. Police were telling everyone after the war that the trouble wasn't over yet and that there was another great disaster pending. Maybe it was the legend of the Shawnee prophet? I remember the Dybbuk Box was attracting people's attention. But they put something real nasty on us!" Dennis's father says. Dennis listens intensely for a moment before interrupting his father. "Okay, Dad. But can you remember Sir Isaac Newton's occult predictions for 2034 or 3060 or anything about 1974 government articles?" he discerningly asks.

Dennis's father remains quiet for a moment. After a few seconds, he responds, saying, "Well, the Nostradamus prediction was aired a lot. The British people have cautioned us about Guy Fawkes Day, the British 5/11, and an 'Evil Blonde Druid priest.' But I recall something about a Medusa or a 'Sighted blonde Dybbuk' that will lead the 'City of demons,' or LA into this great misfortune! I'll give Charles and some of my other old friends a call. Charles was an army field officer in the war. Now he lives in Enfield, Connecticut. He used to write books on the Romanov family and what he witnessed during the war. He worked closely with the Soviets and the old Russian royalty.

"I do recall him saying that after World War II, government officials like the Kennedys held secret meetings. Some of these officials went to Haiti in the late 1940s and 1950s and told the Haitian government and people to stop the practice of Voodoo and Santeria. I believe that Charles should still have some of the paperwork on that. I believe he said that their purpose was to eliminate the threat of another witchcraft attack.

"It is said that John F. Kennedy and other top government officials started to comprehend Voodoo and Santeria.'

"Witchcraft was a topic in many cities of the 1950s. Books about genies, jinn, or vanishing people as characters were all the rage in most cities. These shows posed questions like 'Were there aliens' or Annunaki? There were respectable pilots who had seen things while flying. Were aliens really the insect people? Officials said that these supernatural entities could use flying discs to travel through time. "Who were the insect people that the ancient Egyptians wanted to breed? Were Egypt's insect people with bulging eyes now the demons here to destroy us? Some Greek families still say, 'Bring them back down,' referring to the flying craft. Did what these ancients built actually fly? Subjects like the insect people were the topic of conversation in many US cities of the 1950s.

"The Haitian government responded to the US request by saying they didn't condone the use of witchcraft on anybody. But some of their religions use ancient Egyptian, Greek, and Middle Eastern oracles. Pan, Artemis, Minerva, and Mercury are worshipped in Haiti as demonic gods.

"When I used to study Haiti, I found some curious details. Some people believe in the old Paneas demons from Israel. You should consult your computer about the notorious 'Cult of Pan' in the sacred Jewish city of Caesarea Philippi, Israel. Some believe protective demon guards for the old Turkish government named Hatti are still in control. People have whispered that those demons are very destructive, are uncontrollable, and will attack and destroy anything, even their own people. They still have certain religious followers throughout the Middle East in present-day Turkey, Syria, Lebanon, and Israel.

"Some followers of the occult believe that the 'Antichrist Genies' or the 18 Devils are authentic. Some still believe in the Ptolemy's and Berenice's cult and their time machine? Ptolemy's genies have the celestial number 777777. Many still pray for rewards with these genies.

"Some Haitians have always gossiped those demons are still under their soil and that someone has opened up a black hole in Haiti for the crusted ones to come up through. In any event, the Haitians agreed that they would notify the US government if Voodoo or Santeria was used in a bad way.

"When the Americans met the Soviet troops. The Soviet troops told the American soldiers weird stories like Paul Ogorzow, the S-Bahn Murderer. Some elite German officers often whispered, "I'm Sighted," and not from this time. The troops reported the Jewish prisoners as saying, 'The wicked ones are here, with the snake- woman and the Dodonas. Was the snaked-woman Hydra? Were these Hydra's demonic heads or the serial killers from the Biblical Land of Nod? If anyone calls themselves the Dodonas, seek police help immediately!'

"As the Soviet troops encroached on Berlin, they reported that Adolf Hitler acted weird at times. He saw and talked to a young boy, Sammy, from the United States. Soviet troops reported that Sammy would explain things and see these real demons. Hitler was believed to be a 'Sighted Dybbuk' who could hear and have conversations with Sammy. The Nazi Party, and many Europeans, had a file for the boy and the kids in the future. That's what Auschwitz and Ravenbruk guards said. At least, I think his name was Sammy. Some of the prisoners used to tremble when they heard the name, Sammy. Adolf Hitler talked about him all the time. People used to believe that Sammy was part of the SS.

"When US troops opened the Jewish concentration camps, the Jewish captives foretold of another great disaster coming to the cities of the United States. Some of the German guards and Polish prisoners reported, 'The wicked ones will kill you with viruses next time.' Some of the Jewish victims said that Hitler's field officers had glowing red eyes. And the field officers with glowing red eyes were the ones who used to say, 'We're real demons. These demons are too much for Sammy to handle; let the black manhandle them!'"

Continuing his discussion with his father, Dennis hears his girlfriend, Gloria, coming through the back door from the garage. He knows he must end the conversation with his father and organize things with Gloria. As his father continues the topic of Hitler's Germany, Dennis interjects, saying, "I can't believe my ears, Dad! You sound more and more like Harvey as you discuss Hitler's Germany. You sound as if you're describing a science fiction movie rather than a real war. I'm sorry, but I need to get prepared for tomorrow's classes so I can get some sleep tonight. And Gloria just came home. Can you tell Mom that I'll be heading down to the plantation house this weekend for vacation? The both of you are very welcome to come with us, as you did two years ago, and be our tour guides. I remember both of you enjoyed visiting the historic section of the plantation. Brother Jerry wants time off, and he and I want to make this a joint family vacation." While Dennis is speaking, Gloria passes by him.

Dennis's father giggles and quickly responds: "She knows it's June and vacation season. I'll ask her after she gets back from the grocery store tomorrow morning. Since she's hurt her leg, she doesn't help at the library as much. More than likely, she'll love to ride back to the family home in Olney to visit her family." Dennis's father chuckles as he says goodbye.

After Dennis hangs up his cell phone, he quickly gets up and walks into the kitchen to be with Gloria. While Gloria is preparing herself something to eat, Dennis is the first to talk about his workday: "So, I met this new teacher from LA who says he is trying to understand what really happened with the '08 election, which was like many of the others. Did Hollywood studios make predictions of the future for the United States since the 1940s and through to 2009? He claims that there is some type of manifestation or Antichrist, which threatens America's future. The United States is predicted to be the main country that will destroy the world in 33?6! This destruction may come from natural environmental calamities such as earthquakes or tornadoes.

"Moreover, our plans for a vacation next week are all set. But I'll talk to you in the morning," Dennis says as he quietly gives his girlfriend a kiss on the cheek.

Gloria has been sitting at the table chewing her food while listening to Dennis. Quickly returning the kiss, she says, "I know it's been a long day. I've already made plans for our vacation at the end of the week. I'll see you in the morning. Good night."

Dennis gives her a smile and heads upstairs to their bedroom. Once upstairs, he prepares for bed and goes right to sleep.

In the morning, Gloria, as always, is the first one out of the house. She's usually gone before Dennis wakes up. This leaves Dennis to prepare his own breakfast or else get breakfast at the nearby corner store. After he awakes, Dennis gets dressed and then heads downstairs for something to eat. He instantly searches through the refrigerator, puts his breakfast together, and then heads for the garage. Once outside, he gets in his car and drives to class.

Chapter 2

Theodore and Obama — Presidents of the Unseen World

Bring the Omega and the omicron Genies
(The Reflection: 1902 – 2019. Sorcery in the Chinese Zodiac years of the snake:

1917, 1929, 1941, 1953, 1965, 2001)

Did the Dybbuk in the White Dress of Pittsfield, Massachusetts, Threaten Theodore Roosevelt with the Curse of the US Presidents? Were Barack Obama and Ptolemy Sent to Understand These Hexes and Their Demons?

About ten minutes from his house, Dennis receives an odd call from his father. Dennis quickly pulls over and answers his cell phone.

"Yes, Dad, is everything all right?" Dennis, with concern, asks. "I'm on my way to class now. But I'm only five minutes away from your house."

His father quickly responds, "I called Charles last night, and he told me that you must study Italian and Greek history on 'Presidents of the Unseen World, and his Genies' to understand the curse, Obama, and Head of State Trump.

"But Franklin Roosevelt, his staff, and government officials had always shared stories of the witches' coven cults that follow the current US presidents. His cousin

Theodore Roosevelt had a terrifying encounter with such a group on September 3, 1902, in Pittsfield, Massachusetts. Presidents, staff, and the FBI still have always whispered about one of the most persistent and scariest witch cults exposed. As Teddy Roosevelt was giving his speech, a young white woman of about thirty-five began shouting, 'I want to kill you! Remember Britain's 5/11? Ahh, we put somethin' on you. The big dark head is Hydra's male Medusa. Look what you did to Tecumseh's group! He still lives and will never die! You whites can't understand!

You call us dybbuks, psychics, or witches, but I'm an evil genie from the 'Third Dynasty!' The young woman stood out of the crowd because of her white dress and wide-brim Victorian hat.

"I downloaded one of the Ken Burns PBS tapes of Pres. Teddy Roosevelt that Charles had emailed me. The CD doesn't have sound, but it does have moving images. And you can clearly see them yelling at each other. Old Teddy stopped his discussion with the audience and yelled back at her, *'You better not be the same woman who threatened President McKinley. Dybbuk, leave me alone!*

Dybbuk, leave me alone!' Cautiously, the young woman backed away and looked at the cameraman. Immediately after that verbal exchange, Teddy stopped his speech, got into the chauffeured president's vehicle, and drove off. And within minutes, a streetcar almost killed him in an unusual car accident!

"After the bizarre streetcar accident, Teddy began to believe in the curse, ...jeepers, it was the Chinese zodiac year of the stalking tiger! Remember, he saw President Lincoln's funeral procession. There were other strange assaults against Theodore Roosevelt. Family members have always wondered about the second attempt on his life on October 14, 1912. Government officials had whispered that his attacker heard weird voices about a Mexican 2012! And what force led Teddy to the deadly Amazon in Brazil? That trip almost killed him.

"I have stated that demons are attacking public officials. We must remember England's 5/11 threat of 'Add the Man in the 40's to 5/11!' This warning came after the September 6, 1901, President McKinley shooting. The curse has stated that the demons will kill, cripple, and make officials into buffoons. And that a 'black man and a Ptolemy' will be sent to the Americas to make judgments on the curse. Both the judges are to be high priests. But do the Ken Burns PBS tapes talk about weird demons around Teddy and the government? Clearly, the 9/11 threat came long before any current terrorist activity.

"I believe that the same self-proclaimed group of witches who'd threatened other disasters after they'd threatened President McKinley. Was Princess Diana of Wales's funeral date of 9/6/97 predicted with the McKinley assassination?

There's proof of unnatural death for the Russian Emperor's Paul I in his 1801 letter. But Princess Diana of Wales's funeral occurred 96 years after President McKinley's assassination. To many occult followers, King James VI's '5/11 curse' said that demons threatened British princesses in 1597! And the numbers {96966} means the Prologue to a catastrophe. Back in the 1970's most British children were told to remember September 6!

The Gatchina Letter at St. Petersburg, Russia, was read to Tsar Nicholas II in 1901. This document should be examined because the Russian Romanov family had said that they'd be sending people who'd be explaining the 'weird deaths' of politicians. Historians should reanalyze former President Benjamin Harrison death on 3/13/1901. He was related to president William Henry Harrison and that's where Tecumseh's curse became apparent! Those old Shawnee warriors wanted to destroy the Harrison family!

They should also analyze the Tupac Shakur shooting on September 7, 1996, and the Christopher Wallace shooting. Were those assassinations purposely committed to chasing away people who knew about the deadly 2012 Aztec Curse or the predictions of Princess Diana of Wales and President McKinley? Mr. Donald Trump warned people of the Late princess's premature death on the Oprah Winfrey show in 1988!

Ahh, yes, that group called themselves the 18 Devils or the 'Antichrist Genies,' and they still exist today! But they currently operate under other names. The US government has said they make a political stance at the beginning of the century and in the ninth month because that's the sign of the '18 Devils', or [1+8= 9]! That's where I'd start the research on the Comanche Curse, the peyote rituals, or the New York tragedy of 9/11/01!

This may explain the major world tragedies which occurred in the Chinese zodiac year of the snake? Remember, the snake is the sixth animal used in the ancient Chinese and Babylonian occult practices. Some of these years include (1) WWI, in 1917, (2) 1929 stock market crash, (3) W.W.II, in 1941, (4) 1953 Beginning of Vietnam and ending of Korean war, (5) Nostradamus prediction of 2001.

"But residents around Pittsfield and Lenox, Massachusetts, have said that the Dybbuk in the white dress was a local woman. Her Irish Catholic family had some wealth, but she had some weird fascination with sorcery. The young white woman didn't have any reason to hate the president over Native American issues, but she had confessed to having out-of-body experiences and being sighted. They believed she could foretell the future. Neighbors used her as a remote viewer as she taught witchcraft with about five other women. The one dressed in white was the leader, and her second-in-command was a lady who nicknamed herself Magna the Oracle. The woman nicknamed Magna was friendly and seemed playful with the children.

There are a lot of tales around the group. Because they used to teach the kids how to use demons like Mercury and Jupiter in the old style of Native American black magic! The old-style uses the Babylonian, Egyptian, and Sumerian presidential calendars to maneuver political figures. Some of these Native religious rituals are like the Asian Fung Shui in which they manipulate the Asian calendar. Some say if inaugurated in the 'Year of the rat,' you'll become somewhat scandalous! But if these threats of viruses were made in 1902, then they should become apparent in the year 2019 because of the year's mirror reflections. In some of the Peyote rituals, they've used mirrors to manipulate the medusas and her 18 devils. The reflections of 2019 are clearly shown in their ceremonies!

But Massachusetts State Police want to be notified if these scary demonic witches return to the area. Some say they moved westward. I have Charles on my other business phone. I'll be here all day if you want to stop by later. I have Charles on my other business phone now." Dennis Sr. waits for his son's response.

Dennis Jr. pauses for a minute and then says, "Well, okay, I'll be right over. I'll take the day off since my vacation is next week. They have a trained substitute already there at the school. But what I need to know is how is Theodore Roosevelt connected to President Obama? If you can, ask Charles to find that on CD for me. I'll see you in a few minutes."

Dennis Jr. drives off to his father's house. Within fifteen minutes, he pulls into his father's driveway and walks into the house. Once in the house, he sees his father seated on his couch in the living room. The television is turned down low, and he's on the speakerphone with Charles. Dennis walks over and gives his father a strong handshake.

"So, this is the Teddy Roosevelt CD? Does it show FDR and other presidents? It must have camera footage of European royalty," Dennis Jr. says after taking a seat beside his father on the couch.

His father scratches his head and says, "Ahh, I was just watching the part about when Teddy Roosevelt passed through Khartoum, Sudan, Kenya, and parts of Lower Nile. When he was young, he spent some time in Egypt. And history has always shown Teddy's big hunting trips through British Southeast Africa. Charles told me he was not just after trophy animals but answered about witchcraft. If you notice him asking tribal people, 'What's moving around in me, and how do we get these "spirits" out of us? They are still killing you people, and now they're killing Americans! And what does Alkebulan, Land of the genies mean?' There are some historians and Kenyans who believe the 1902 streetcar accident spooked old Teddy. Remember Teddy and his staff were the first government officials to report Unidentified Flying Objects! Maybe, government officials believed in the Tecumseh and the Pittsfield genie's curse on the US presidents and its people.

"Charles had just told me that Washington DC withheld many secrets about the predictions of a 'black man president around 2009' and a Ptolemy from Egypt. But Charles had advised me to watch *The New Soupy Sales Show* circa 1974. Those shows may have exposed a lot of the worries the Kennedys had about these manifestations, Germany's demons, and the threats to his family. Those CDs may show the psychic prediction that only the black man or high priest would be able to assassinate specter Osama.

"Government families like the Kennedys of the '50s and '60s really put their fears and warnings about these manifestations and their demons in *The New Soupy Sales Show* circa 1974. I believe they whisper about the nightly attacks on Ptolemy's-Eye and people of the future. These manifestations would expect Ptolemy relative to warn the municipalities about what he sees with his eye of Horoscopes. Let the people know that demons exist and are not just myths. The German people warned future governments that these manifestations would spread viruses to cripple the people as they did to Franklin D. Roosevelt with the poliomyelitis virus. These demons will try to provoke the public and the US government to attack each other. Remember Napoleon's army whispered, 'Tharu or Rhinocolura's demons and the Incubuses have made us destroy our governments.'

"I have to recheck Teddy's CD in the Wilson administration era, but was someone brought over from Alexandria, Egypt, or Greece to help explain the Antichrist Genies? Were Berenice IV's good genies or an ethnic 'Black Man' to follow the Ptah later? Were there to be two 'Presidents of the Unseen World' to help explain the Nostradamus prediction? One with the 'Eye of Horus' and one without the eye?

"Did Teddy Roosevelt send orders to find someone from the old Ptolemaic family back in the 1920s? Charles had said that a Greek royal family had a lineage that dates back before 356 BC. If you read Daniel Chapter 2 and Genesis 11:1-9, you can understand Ptolemy's and other warnings of collapse of a giant tower or statue by daemons. You can study the Antichrist genies and Genesis 9/11 in some Theologian classes. These daemons would try to end the world 1366 years from 2000, or 33?6!

Along with Aesop's fables, there used to be a book on sorcery and witchcraft called the book of Nod. There have been many children's books about the fabled People of Nod who could vanish. But the ancient book Nod's sorcery was written about in the bible! Today, many Sicilians share passages from those ancient books. Tales of electricity, flying objects, and lights without heat in Alexandria, Egypt, and Sudan. The Italian government asked the Africans to bring electricity and the steam engine to Sicily and Rome two thousand years before Edison's lightbulb and automobiles!

The cult of Janus is still practiced today in certain Italian towns. Ancient fortune tellers have always discussed the activities of the 20 medusas in the 21st century! These ancient Greek families used to study with the First Dynasty in Memphis, Egypt. And we should try and save the Ptah because we desperately need his eyes and ears. The Ptah lineage was like a Jesus to the early Egyptians, Palestinians, and Jews. Anna Roosevelt may've met the white man from Egypt or Greece. I must analyze her conversation with her parents. I've seen Tsar Nicholas II of Russia at play with his Greek relatives in their home videos. You can visit Vergennes, Vermont, and find out more about Ptolemy-Berenice, genies, and the Children of the Future. They were always talked about in the palaces of Great Britain, Europe, Africa, and the Middle East.

"But did the North American Plains Indians talk about Cleopatra's or Berenice IV's relatives? Books like *Stolen Legacy* by George G. M. James (from page 177) talked about warnings from the Grand Lodge in Thebes, Egypt, to its minor lodges in distant lands. The temple's president and priests had said that their own metaphysical creations were now attacking them! And their demons would be attacking future citizens in the cities of their minor lodges! People in Palestine, India, and North America should analyze my writings about the 'Terror of the Genies'! But did they use Ptolemy-Berenice's eye? Supposedly, this is someone living 150–200 years in the future. Then Sitting Bull's family of the Lakota Sioux should be contacted." Dennis Sr. fast-forwards through Roosevelt's CDs.

Dennis Jr. quietly nods as he watches the television. While making notes, he says,

"Some of those LA actors should explain the definition of a dybbuk. But few don't understand that the early Egyptians created people who'd become clairvoyants, sighted dybbuks, or possessions. Some believe that LA, California, was nicknamed 'City of the Genies, or Demons.' The Polish and the Germans believed they had good and bad ones. Why did the American government label both Teddy's heckler and the psychics of the concentration camps as dybbuks? Didn't male or female dybbuks start up the old Shaker and Quaker groups in England? They were used to accepting the idea of Artemis and other spirits making your body shake or quake. Are these the manifestations that are channeled in the biography of Winston Churchill, *Hero of the Empire*, written by Candice Millard, on page 25? But you should always check Pope John Paul II's trip to Ireland. 'Things are Humping Priests around! ... Please Bring Ptolemy-Berenice!'

"The public needs to know about the Pittsfield dybbuk's curse, Britain's 5/11, and the 9/11 prediction. Were Mr. Obama and Berenice IV's relatives sent to judge the curse on the presidents? I must reread the prophet's curse and check everything. An Egyptian High Priest is initiated at age thirty-four in order to become president! They should recheck the Dodona's Priestesses birthday in 2008. They have to be some of the lead sorceresses in this kind of magic. It's very unfortunate that sorceress Artemis has been penetrated by the Succubus genie.

"Dad, you might prove Harvey's point. Something may be wrong with Hollywood's Christians, Scientologists, and the government. How could an American Plains Indian know anything about Cleopatra, Ptolemy, or Memphis, Egypt's Third Dynasty? I don't mean to underestimate anyone, but I didn't understand Egypt or the Ptolemies before college. I went through seven years of world history before I could follow Berenice IV and ancient Roman-Greco artifacts." Dennis Jr. is still fumbling through his father's CDs.

As Dennis's father ends his conversation with Charles on the speakerphone, he begins to speed-dial another number. Within minutes, Dennis Jr. begins to hear the people from the Onley, Virginia, plantation.

Dennis Sr. turns to his son and says, "I'd like to introduce you to Franklin, one of the employees from the Onley, Virginia, plantation. He knows a lot about ancient, mystic Native American and African cultures. But the old Roosevelt tapes show presidential visits to real tribal Indian villages back in the early 1900s and the end of the Indian Wars. Maybe you should talk to Franklin, and he could tell you more. Franklin is on the phone now, so all you have to do is talk into the speakerphone." Dennis Sr. begins to sit quietly.

Franklin in Onley, Virginia, breaks the silence first and says, "I've heard you want some hidden facts about some Native American history and legends.

Unfortunately, some believe the male Medusa was used on the Native Americans? These are the demons and evil spirits of Venus, which destroyed the Azteca of Tenochtitlan, Mexico. And now, this demonic force is controlling the Shawnee Prophet and his Jinns. You must read Edward S. Curtis's '*The Indians of Oklahoma. The Wichita. The Southern Cheyenne, Oto, Comanche, and the Peyote Cult*. There are discreet passages that discuss the Comanche peyote rituals and more!

"But these demons are being led by the Incubus Prototype and are destroying everyone. But our war is not with anything human but an unseen predator, the Incubus Prototypes. Humans designed those things. But now, Medusa and her demons are like uncontrollable robots. And its demons are designed to turn average men into serial killers! There's usually a dybbuk woman who calls herself Minerva. She is reluctantly used as the eyes and ears of this 'great Satan, and the Antichrist Genies'! She will help you locate them.

"There have been conversations about Native American tribes using sighted dybbuks. A dybbuk is an Eastern European name for a bad possession. But research has shown dybbuks differ in their functional abilities. Is there a difference between a Victorian-sighted dybbuk with the ability of remote viewing and a good psychic in Hollywood, California, in 2002? Maybe, you can find the answers about the dangerous 33rd legion? But in East Africa, the study of genies, both negative and positive, has become a science referred to as "genism." And there are different types of genies.

"Some philosophers believe that most people have entity or Jinn within their body. Just get a common household Ouija board and sit in a quiet place by yourself. Eventually, with only your hands on the slider, the needle will move by itself without your control. But a sighted dybbuk could have many more entities and is considered dangerous. Regrettably, there's no way to control what flies in and out of your body. The 18 Devils with their 2400/55 apocalypse are what has destroyed families and their religion. That's why you're told never to continue channeling each other.

"I don't know everything, but maybe you should contact Sitting Bull's descendants. Some believe that Lakota Indians used Indian black magic and the male Medusa on General Custer in the Battle of Little Bighorn. Certain practitioners claim that early forms of Indian black magic include calling certain genies and the insect people from Egypt's Second & Third Dynasty. The line of Cleopatra and the Berenices were also contacted because they used the hierogrammatic priest. The hierogrammatic was the priest who sipped those libations of human blood and colored photons. Their priests were reported to pass through time with their obelisk guardian demons. There were about twenty or more ancient Egyptian priests who used hypnosis and demonic Incubus Jinns to support the seven-foot male human head. These are the priests who claim to make their subjects' eyes glow red as they cross through time as photons. Remember, you must understand time, not space, to understand the old texts.

"Some southeastern Indian tribes from Memphis, Tennessee, to the Black Hills in Wyoming claim that there's a time portal in the central part of this nation. There's two-hundred-year-old folklore of an Iowan tribal members. Many spoke of an ancient tales of King Nebra in the north. King Nebra guarded Native Americans and whites. He would feed people and the animals in the winter. White settlers in the 1800s also felt the King Nebra spirit would help them. Maybe this is coincidental, but there was a King Nebra in the 2nd. Dynasty in Memphis, Egypt.

King Raneb was worshipped as a god in the 2nd and 3rd. Dynasties. But how did the US Native Americans know about an ancient Egyptian King?

"In certain areas around West Memphis, Arkansas, Native American folktales still exist, some witnesses have said, telling of those who have spiritually summoned different gods and demons. And spirits had inhabited their bodies when they ghost-danced and held other rituals. Even today, if you meditate or sleep in these areas, you might hear or be touched by the Shawnee prophet. In these West Memphis, Arkansas, areas, Ouija boards act in unexplainable ways."

Dennis Sr. and his son listen patiently to Franklin as he tries explaining his knowledge of Native American black magic and culture. Then turning to the speakerphone, Dennis Sr. interjects, "Yes, I was always suspicious of the Ghost Dance. The Union soldiers used to group all the Native Americans together, make them dance, and shoot off their guns. The ones who had become warlike were destroyed. But does Egyptian history show any interaction with Native Americans? Didn't the Wilson administration bring someone over from Greece or Egypt to help explain the Native American prediction of an American 5/11 in 2001? Is this the Moses or the high priest in the biblical verses? Was this Berenice IV's genies, and the connection to Banias, or Caesarea Philippi, of Palestine?"

Franklin quickly replies, "Maybe they were writing about the Nostradamus prediction and black magic. Were citizens of Memphis, Egypt, was put into a magical sleep because they didn't want the predictions of a 2400/56 apocalypse of slavery to occur? Some Jewish Kabbalah followers believe that every 2,400 years, there's a 6+6+6+6 civilization apocalypse! The terminology is written as a Lambda Calculus equation of {24/56 PY}! The smaller wars and disasters are written as 5/11, 9/11, 6/11, and 6/11. Within these smaller wars you should be hauled away by the 18 devils and will lead to a 23/?6 civilization apocalypse like slavery. Some rationalize that the 23/59 apocalypse prediction also happened to the Native Americans from 1518 to 1815 and 66 years after. A small 22/59 apocalypse even happened in medieval Europe. We'll explain more about the Antichrist genies later.

"Tecumseh's people have always written about 'Judgment' coming and pushing evil demons away. Middle Eastern history talks about a trio of genies who call themselves Judgment who follow the Ptah. Some of these genies were the chief justice or president of the unseen world. Information on: The New Caesar, Ptah, and their genies, read *Stolen Legacy* by George G. M. James, on page 107.

"Will the Medusas or the Great Satan lead you into self-destruction? They've set up a twenty-four-hour prayer system so those sighted demons won't return.

"Was the American and the British government searching for answers around Meroe, Sudan? Are these metaphysical people or Incubuses written about in Akhenaten hieroglyphics? Was Theodore Roosevelt perhaps looking for answers for pesky dybbuk witches or genies?'

"Well, I heard you'll be coming to Onley, Virginia, next week. I'll try to have more proof next week. Thanks. You've been supportive of my conversation." The speakerphone conversation is ended.

Dennis and his father view the CDs of Theodore Roosevelt for a couple more hours before Dennis Jr. gets ready to leave. As Dennis puts the tapes into his briefcase, he acknowledges his father and says, "Thanks, Dad. I'll have to read what Winston Churchill thought on manifestations in the early 1900s. I'll give some of these tapes to Harvey for his research. Are there any covens of witches that follow these current US presidents? And what's lurking in the Central Plains, in Los Angeles, or in Las Vegas? The people from the 2017 Mandalay Bay Casino calamity should be contacted. But I must get ready for school tomorrow. My prayer is for someone to check into these incidents because this is not a joke."

Dennis gathers up notes. Within minutes, he hops into his car and heads for home.

Chapter 3

Nightmares of Crumbling Towers or Metal Statue?

Decoding Sir Isaac Newton's documents on the 21st Century. Daniel Chapter 2, and Genesis 11: 9. The 18 yr. disease curse of Exodus 9-11? Bezalel's 65 Legions of Devils. Hercules and the Prince of Peace. Minerva (Goddess of War) and other supernatural folklore.

For Dennis, the next three days of classes pass quickly. And when Friday, the final day of the school year, arrives, all the teachers are prepared for their vacation time. Harvey brings his laptop computer and seven DVDs about the predictions into the teachers' lounge. While the teachers take lunch, they view Harvey's DVDs. Most of the same teachers are here today who were here the other day. A few have brought some of their friends.

Two of the DVDs contain recordings and writings of presidential reports from1974. Along with the taped recordings are written manuscripts by Sir Isaac Newton on Daniel Chapter 2 and the metal statue. That giant falling metal statue was believed to be New York's Statue of Liberty, or the World Trade Center's! Middle

Eastern peoples and citizens of Nineveh, Iraq, have always warned people that the Antichrist Genies would come and destroy them 2,400 years later! Ancient people used warn us about Genesis 11:9, Hydra, and the eleventh genies. They didn't want all these wars in Iraq. People still remembered the warnings of the dangerous Succubus of Mesopotamia, especially after the demolished Thebes,

Egypt, in 564 AD. In occult theology Thebes, Egypt had a catastrophe in the year 2300/564. Was this a prologue to the Gulf War or the Iraqi Civil war?

Some New York citizens in the nineteenth century protested the building of the Statue of Liberty within those nine years. Occult followers believed that Beelzebub would use the 33rd. Legion and followers of the Thirteenth Angel to destroy the metal monument or the Towers of Babel with the British 5/11 curse or the 2012 Aztec curse! Within approximately five years before 2012 the Ninth angels or Bardiel's followers would help rebuild these monuments! These are the seven devils described in the Dresden Codex. Even today certain Kabbalah followers use the Exodus 9-11 curse to chase away the evil magicians and other hostile forces with viruses!

But what did you think of the Russian Emperor Tsar Paul I, letters in 1798? Were there warnings of demons before his assassination in 1801? Did the 00' curse begin with the Romanov family? While the teachers view the DVDs, they see some comedy, but they observe more threats of a manifestation in the year 2009. While Dennis finds the writings of Tsar Paul I or 'Mad' Paul enthralling! And Grigori Rasputin's thoughts on Russia's demons and predictions span well into 2018! Dennis knows he'll have to provide others with this information!

As the men watch the DVDs, they quietly compare notes with each other. Dennis is the first to speak out: "Okay, I just saw them talk about letting the black man assassinate specter Osama. And even an assassination date of six months before 2012! ... Well, the deadly Aztec curse of 2012 is pinpointed! But it's hard to tell with all these religious wars. There's always been more than one religion practiced and caused controversy. Rumors around certain Middle Eastern communities state that the Baal god and his 66 Legion have returned with its 'Caged Eye Pentagram'! But the 'Succubus of Babylon' controls the two 33rd. Legions, Baal doesn't really exist. I'd have to research the 'Caged Eye Pentagram.' Does it bring the antichrist?

But first we must consider the similarities of the Jewish ceremonies of the B'nei Mitzvah to other worshipers of the 13th angel. These celebrations of the 13th year to adult hood exist in Kabbalah with their belief in the 13th angel and within the Mesoamerican gods of the 13th Baktun? Is there a connection between the gods of Baktun 13th and Bezaliel's 66 legions? How does Bezaliel, or the 13th angel, control his 200 watchers in the Book of Enoch? If you study these gods of Baktun 13th there's apparently a weird shift every 400 years!

But these tapes appear to be discussing a scandal that will disqualify someone from becoming president or president-elect. The tapes suggest that there is someone practicing sorcery in the United States. And the tapes suggest that the 9/11 tragedy was organized to happen on the eve of the autumnal equinox or Halloween. Maybe sometime during the three months of September through November. That's when the witches used to come out of the body and observe the transition into the bad weather and into deadly winter. Is there a connection between the 'Antichrist Genies' and the connection to WWI's end date of 11/11/18? ... Why are there a lot of murderous wars or viruses every eighteen years? Some believe that's the occult sign of the Antichrist Genies or the 18 Devils.

These recorded tapes suggest that Barak Obama was supposed to control the Antichrist genies in 2009. Did medieval peoples try to control the 18 Devils when they stopped fighting with the Holy Land's ninth crusade in 1271-1272? People have always said that those demons would be back with another attack in the Holy Land within seven hundred years! Were the World Trade Centers built in 1971 to bring attention to these demons and rein peace among the nations?

Franklin gave me details about Israel's Fourth Crusade in 1199-1204. The crusade has the occult number of 18/11. Franklin had informed me that 11 divided 18 =.6/11. So, realistically .6/11 is the first religious war in the 23/56 apocalypse series, and 9/11/01 will make the third. Legend states that the third religious war in the Holy Land should bring the Holy Trinity.

"I see that the tapes explain the stereotypical witch riding an old shaman broomstick. The black hat means the person who wears it has the demonic black magic knowledge of the old pyramids. The witch's characteristic black hat is the symbol of a pyramid. It is reported that the old priests used the formula Time divided by Speed =Distance. Did they actually fly through time to prepare for another year into the future on the winter equinox holiday?

"Unbelievable! These tapes talk about what Comanche Indians had thought about Ptolemy's genies, Obama, Trump, and the 9/11 attacks! And what about threats to Victoria's family and to the new city? Also, a young man named Sammy is both used as a scapegoat and threatened with a virus. A famous movie producer is warned about an evil blonde giving a virus to the people of LA and other cities. These tapes explain that she is the same evil blonde who calls herself 'Atom's Wife' or a small Medusa. But are there warnings of threats to our presentday society in these videos?" Dennis gestures to Harvey.

Harvey's eyes open wide. Giving an immediate response, he says, "Now, I told you that you may hear some eye-popping conversations from the past. But we must investigate the dangerous '9 Demons' on Queen Victoria's family! Queen Victoria had nine children. And the Nine dangerous demons who'd cause 9/11. In the Kabballah cult, there's nothing after nine. And those nine demons will try to destroy New York, America, and everyone!

Have you heard about the diseases within 18 years after 9/11? But you must read Sir Isaac Newton's predictions for 2034 and 2060. The Ptolemies were supposed to send somebody who could see this dangerous succubus. But there're written notes from 1718 describing demons who'd caused Alexei Petrovitch's execution by his own father.

Just compare the similarities between the Russian and American governments between 1718-2018. The Russian letters clearly discuss the presidential diplomats being sent to the Americas. Was Sir Isaac Newton trying to explain the 33rd. legion with this equation: 2034- 33rd. legion= 2001? Are there victims of the 33rd. legion every 34 years? If you have questions or feel threatened in any way, feel free to complain," Harvey says, viewing the DVD on his laptop.

The men watch some of the remaining videos for about half an hour before they clean up their classrooms and go home. All their students have gone home by now. Since this is the last day of the school session, there was only a half-day for the students.

Before the teachers leave, Harvey reluctantly asks for his own videos back. All the men complain and ask Dennis to make them copies. He patiently agrees to the guys' requests and thanks Harvey, going back to his classroom.

Since this was a half-day, there wasn't much to clean up. After an hour of paperwork, Dennis locks the classroom door and walks out to the teachers' parking lot. Once he's in his car, he calls his girlfriend and then heads for his home. Within the hour, Dennis parks his car in his driveway, grabs his briefcase and the bag of DVDs, and heads for the back door off the kitchen. Once he's in the kitchen, he turns the lights on and looks around the room for his personal computer. Unable to find it, he places his briefcase on one of the kitchen chairs and takes his bag of DVDs over to his living room television.

He immediately places one DVD inside the player and then goes back into the kitchen to prepare himself something to eat. After he grabs a sandwich and a glass of water, he walks back into the living room and gets comfortable in his reclining chair. With the remote in his hand, he quietly searches through his DVD that is playing on the television. After a couple of hours of watching different DVDs, he starts to doze off in his lounge chair.

Gloria comes home about three hours later. When she enters the house through the kitchen door, she finds her boyfriend asleep in front of the television. Trying not to wake him, she immediately goes upstairs to find a blanket.

By the time she comes back downstairs, Dennis is somewhat conscious. As he wiggles around in his lounge chair, he says, "Yes, Gloria, I know you want to get something to eat, but Harvey brought his tapes in today. I want you to look at them when you get the chance. They suggest that the Kennedy administration spoke of a manifestation and its demons threatening America. These tapes suggest that someone is using magic on everyone. Were these the ancient

Mesoamerica sorcerers from Palenque predicted to arrive sixty-six years before 2012? Did President Kennedy call for somebody when he was being shot at? And there were reports of weirdness in the 2008 election," Dennis says, sounding suspicious as he watches the TV.

Gloria, giving Dennis a blanket, watches the television for a couple of seconds. Then she goes into the kitchen and looks for something to eat. As she rifles through the refrigerator, she says, "After I get my sandwich together, I'll take some of the DVDs upstairs with me. My computer is upstairs. We'll have plenty of time to view the DVDs because my vacation started today. By this time tomorrow, we'll be on our way down to Olney. We'll have to go by your mother's house and pick her up in the morning. But we can watch the DVDs in the car on the way down. I think it's an eight-hour drive."

Gloria comes out of the kitchen, goes over to the TV, rummages through the bag of DVDs, and selects two of them. After she adjusts her food and the DVDs in her hands, she starts to head upstairs. On the way, she turns to Dennis and says,

"Okay, Dennis, I should be up early tomorrow. That was a good decision you made to use your brother's vehicle for our vacation trip. Since Jerry's car is in our garage and fueled, we don't have to waste time or money on a rental. I've already packed our camping equipment and clothes. All we have to do is pick up your mother and go." Gloria waves and heads upstairs.

Dennis smiles and then leans back in his chair and watches his DVDs for about an hour before going to bed.

With the sunrise, Dennis is the first up. There isn't too much packing to do since all the camping equipment is already loaded in his brother's car. Dennis works out. Then he calls his mother to ask what time she wants him and Gloria to come and pick her up.

Apparently, Gloria changes her mind and sleeps late. So, up at nine o'clock, she showers and hurries downstairs to grab something to eat before they go. Dennis is already in the living room dressed and ready to go by the time she begins to eat her breakfast. As she eats, she watches Dennis talking to his brother on his cell phone.

"Yes, Jerry, I will miss you if you don't show up. We'll be down there for about two weeks. I have political things to talk about. We're going over to pick up Mom after Gloria finishes eating.

"I'm not sure, but I think Gloria is about ready to go. I'll talk to you later, "Dennis says as Gloria gestures to him.

Once he hangs up the cell phone, Gloria walks over to him and says, "I'm ready to pick up your mother now. It's best to get an early start. I'll get my jacket and meet you out in the garage." Then she walks back to the door off the kitchen.

After locking the house and arming the alarm, Dennis walks into the garage, gets into his brother's car, backs out into the driveway, waits for Gloria to get in, and drives off. Dennis's childhood house is only about an hour away, in Greenwich. His parents' neighbors respect his parents, who have been married for thirty years and who keep a good, clean house.

Dennis has already called his mother, Sylvia, to alert her that they are on their way, so when they pull up in front of the house, they find Sylvia waiting on her front porch with some of her luggage. Dennis quickly loads up Sylvia's bags as she gets comfortable in the back seat. Dennis's father waves from the front porch as they depart, and then Dennis drives slowly down the street.

Dennis doesn't waste any time in heading down I-87 South. This trip reminds him of a childhood tradition that he and his siblings had of meeting up at the historic house in Virginia. He wants to call his younger brother, Jerry, but he knows that Jerry is still working and will be down sometime later this week.

They are almost in Albany, New York when Sylvia breaks the silence and discusses her former job of being a tour guide. "Did you know that during the 1850s, many abolitionists took this same path down to our historic plantation in Olney?

Sometimes they were led by some well-known abolitionists, like John Brown and Catherine Keese, and other times they were led by lesser-known abolitionists, like Artemis, Dodona, and a woman who went by the nickname of Minerva. Yes, sir, it's been written that Artemis and Dodona, both of whom taught magic to the Amish, were reported to have said, "We are not from this time!"

"Many of the Amish had talked about a 'weird future' of Berenice IV's relative with the Eye of Horus or the President of the Unseen World. And if you presently visit Poughkeepsie, New York, they still talk about Artemis's genies and the Dodona. But one of the major plantation folklores is the legend of the Aaliyah genie! The Aaliyah genie was always discussed in the Italian Villanovan culture and throughout the Middle East. This genie has always warned people about the dangerous Babylonian succubus which destroyed Armenia, or biblical Nod. What does the term Fallen Angels mean to many occult followers? I wanted to investigate Kobe Bryant and Aaliyah, the musician's strange death on 8/25/01. I have proof that Kobe Bryant's premature demise was predicted a couple years before 2019!

And what really happened with Ms. Lisa Lopes' car accident on 4/25/02? Did John Belushi predict radio DJ Ms. Martinez's car accident on 11/4/19, on the eve of Britain's Guy Fawkes Day? And why did that accident occur seventeen years after Ms. Lopes's accident? I tried to warn both parties about these predictions since 1997.

But names like Ulysses S. Grant and other Greek goddesses stood out among the population. Some Amish people have said that these white female abolitionists used code names to escape capture.

"But the Amish who worked with and knew them said that these women were possessed by the Greek goddesses after whom they were named. I remember they said that some of those women worked closely with John Brown and Harriet Tubman. If you go to Rutland, Vermont, people there will tell you that the woman nicknamed Dodona disturbed John Brown. John Brown said that she came out of her body and talked to him all night. Some believe that the legendary John Brown was forewarned about his own death. Word was beginning to spread about spirits possessing the captives and the coming of Ptah. Some remembered the Mexican prophecies released to Father Bernardino de Sahagun in the 1580s. They knew of the genies who'd follow the Ptah or Berenice IV's relative in 2012, which would lead to the president of the Unseen World!

These metaphysical people or Jinns would have to introduce themselves or see the Ptah. John Brown had left warnings for the Ptah that certain devils would try to push him and others into a fiery battle against the infidels in the year 2,000. And the deadly blonde Sphinx and the Incubus were leading him to his death.... Back then, "Nobody can understand genies or demons!" And there are other stories of the spooky area too.

"From Minerva, New York, to Poughkeepsie, New York, the historians are keeping some things secret. They used to call this area of the Adirondack Mountains the spooky area. People who knew Artemis said she had out-of-body experiences and would touch people when in this state. And the one who called herself Dodona performed magic and taught it to some of the Amish women. The Amish say that she used to make walking sounds across the floor, but nobody was walking! There'd be a knock on a door, but when you answered it, nobody would be there. On top of this, Artemis said that she and a man nicknamed Mars would come back and teach the Amish children magic too.

"The Amish living around Rutland, Vermont, have always said that if you see any short dark men with glowing red eyes, you should stay alert and give them shelter because they're possessed. The old plantation owners used to call on the Amish to take the possessed slaves up north and hide them there.

"Artemis still has religious followers from Poughkeepsie to Wilton, New York. And they've been waiting for this Ptolemy and his presidential genies, in the year 2,000. All you need to do is ask, 'Where did the possessed white woman live who helped the ex-captives? And where are her dancing genies?' Someone will show you to her house. All this happened sometime around the turn of the century. There's an older woman living on our plantation in Olney who informed me of that historical fact.

"I think I've told you some of the stories before, haven't I, Dennis?" Sylvia asks.

Dennis is driving and listening to his mother's conversation. Responding quickly, he says, "Yes, many times you've mentioned it. I thought the name Minerva rang a bell. There's a folk legend about her in the Saratoga, New York, area. People around our area say that she was a specter who haunted the area around I-87. She looked for fighting men during the American Revolution and the Civil War. And Amish farmers reported that the specter, or sorceress, Minerva sometimes hopped into bed with them. The old Amish men used to say, 'Is that you, Minerva?' She'd say that there was some type of weird time warp over the area!" Sylvia adds, "Marie, the Cheval witch who lives on our fifteen-acre plantation, can tell you more about these people. Your great-grandfather had left her family about two acres of land not far away from the historic mansion. Marie is the one who retains our family history.

"You and Jerry had a ritual of investigating the abandoned quarters, which still stand on the historic tobacco plantation. You both used to be awestruck by the Civil War relics and historic structures around the town," Sylvia says.

Before Dennis has a chance to get a word in, Gloria interjects, saying, "Yeah, yeah, I can't wait to see those old huts either. And you say she still practices Santería or another religion?" She turns and looks at Sylvia.

Sylvia clears her throat and says, "Now, she's older than me. I remember when I was younger, people whispered to me, 'Go to the witch lady nearby. Ask her anything. I bet she knows.' Marie said that her great-great-grandmother used to live in one of those huts. She told me that they used to practice the ancient religion of Santería or voodoo or something like that. Oh boy! You know, I even saw some of those old religious artifacts. She told me that the Cheval witch lived there, and that's why everything is preserved so well. She even told me that the Cheval witch is related to us through the planter. Ulysses S. Grant had visited Marie's great-great-great-grandmother. Our plantation once was a rest stop on the Underground Railroad!"

Still driving, Dennis peers through his rearview mirror at his mother and says, nobody told me anything about all this. You said that this black woman is related to us? We're Greek, Native American, and Arabic. Nevertheless, what's the deal with Ulysses S. Grant's Demons? Why have historians always labeled the general as Ulysses Trojan Horse filled with demons? Oh, she's related to you because your great-grandfather had servants. Nothing more needs to be said!" Dennis returns his eyes to the road.

Quickly seeing her son's interest, Sylvia, folding her arms, says from the back seat,

"I think it's best to let Marie tell you the rest. But you must hear the story of how some of the African captives believed that Ulysses S. Grant was possessed by the Odysseus Jinn. And the ancient sorcerers had always used Ulysses and Hercules Jinn together. But we have about four hours until we get there. I'm going to get a little sleep back here." Sylvia looks out the window.

The car falls into silence. Dennis drives for about two hours and then stops at a rest stop to get some food. After a twenty-minute snack break, Gloria drives the rest of the way to Virginia.

They reach the plantation house at nine o'clock. Plantation workers help them unload their bags and show them their rooms.

The three travelers, saying they are very tired, wave good night to everyone and walk to their rooms. The plantation house is a three-story mansion with many rooms. Dennis, Gloria, and Sylvia have all requested rooms on the second floor. Sylvia was raised in this house, so she takes her childhood room, while Dennis and Gloria take a vacant room down the hall.

Dennis still has many questions. He decides to ask for some information from some of the plantation workers. The mansion's lower rooms are rented out because the house is leased to a few small companies. There are workers downstairs always at night. As Gloria unpacks everything, Dennis tiptoes downstairs. He decides to go to the kitchen, where there is a little cafeteria set up for the workers. He begins to ask around and soon starts getting answers. He finds that one of the plantation workers who grew up in this area, Josh, has answers for him.

Josh says to Dennis, "Yeah, I couldn't believe in the 2012 Aztec curse myself! Historians have reported that a Mayan king, Pakal the Great, in 611 BC, still channeled people in 2012 -21! It's hard to conceive those Aztec priests from the sixteenth century could travel through time!

But these 'Children of the Future' were prophesied to be taught by the vicious 20 medusas and were supposed to teach Ulysses S. Grant, the Navajo nation, and the government about the Anunnaki's weird religion. Grant left word around the plantations that the captives weren't supposed to come out of their bodies during religious practices and that they should entirely cease worshiping demons. And inform government officials about the use of Black magic and the 20 Medusas.

"The Maat Boy was supposed to be placed into the family of the Cheval family at the approach of Europe's New Caesar. The family had come from Alexandria, Egypt. Historians left word that the Maat Boy and the children of the future were supposed to explain the first and second manifestation of the 24/56 apocalypse. But nothing would stop the third disaster from coming. That manifestation would try to kill everybody. The plantation owners didn't like what they were seeing and hearing.

"You don't use Egyptian black magic because you're risking your own life by calling the 18 Devils or Antichrist Genies. Some of the captives believed that General Grant was possessed by the Greek Odysseus Jinn! ... Was he followed by the metaphorical Big Trojan Horse filled with demons? Some believed that the Male Medusa followed General Grant throughout the Civil War and into the Native American Indian wars? People have said that his request for Hercules or Berenice IV's relatives supports that rumor!

"But Marie warned us of Minerva and her cult of Anubis. She had said that this cult of the Annunaki used the 'insect people' and their flying time machine. This is the cult that would bring the evil three and their demons through a time portal in Minerva, New York. The Insect people sound funny, but some Western Native American tribes still channel them in rituals. They used to make metaphysical people out of fire atoms or electrons. These atmospheric electrons are the same things that cause rainbows or the aurora borealis to appear.

"Only deified people should attempt to 'Extend Yourself' or have out-of-body experiences, and that deification is what is meant by eternal life. After going through that step, you can proceed either to the next phase of reincarnation to stay godlike. The priests also used to create little demons that looked like holograms or cartoons. Some of these demons were made with electrons from human brains, so they did have the reasoning capacity of a human, but the old priests just gave them a body. Because they were made too well, they started to control the minds of average people. There is nothing mystical about the creation of demons. We make electrons or demons walk today, but we call it television!

"I should warn you to be careful when you see obelisks or sphinxes in cemeteries.

Because these monuments are considered to be possessed with sacred deities. These deities are believed to have possessed Osama bin Laden and are thought to be controlled by Minerva or the evil blonde genie in the west. Minerva, on the other hand, is a white woman who is reported to be exceptionally beautiful, comparable to a fashion model.

"Minerva travels through time in the oracle's hidden time machine. But her family made a tragic mistake. Everybody has heard of the legend of the gorgeous but deadly Medusa. Don't watch evil Druid priestess dance because she will stun you with her beauty before she kills you.

"The Ptolemy family of Alexandria, Egypt, reportedly could see oracles like Minerva in their electronic state. Ancient rumor has it that Minerva and the Uhuru sorcerer work together. And that the Uhuru sorcerer can split the fabric of time with his hands when he is in his electronic state. These are famous Jinns and genies that many old Comanche and Shawnee tribes used. This is important to remember because you must know what time period the genie came from.

"Some Native Americans used to use technology from these ancient genies. They tried to use the knowledge of the Egyptian scarab's antibodies to make themselves immune to disease.

"Yes, but Sammy's people have the Eye of Horus. This family can see and hear the invisible ones who are in the atomic state. The family of Cleopatra and Berenice IV is written about in the Bible along with King Herod of Palestine! There are still temple remnants of Berenice of Caesarea Philippi in Israel or Palestine. You don't have to be a scholar to understand they used to make genies or Jinns. The old Egyptian priests had developed an eye, and they called that lineage the Demiurge. People still whisper about the genies who follow those old Royal families.

"Oh, I think you may be in luck! Marie is eating here tonight. She's right in the cafeteria's kitchen. I'll get her," Josh says, getting up and heading into the kitchen.

Just then, another field worker, Franklin, walks up, taps Dennis on the shoulder, and says, "Ah, I think I know what you're looking for. My name is Franklin. I hope they're not scaring you too much. I heard you and him discussing this Great Satan, the 18 Devils, or disastrous 9/11 curse. My father and I, both, have worked here for a long time, but I don't like scaring families or anybody else with these tales of the wicked ones or Ulysses' Trojan horse filled with demons. Some believe that the wicked demonic shamans have moved to Los Angeles." Franklin bites into an apple.

Still eating his apple, Franklin sits down at Dennis's table. Then he says, "I don't see anything wrong with bringing that up. It is remarkably interesting. There were many folktales about demons and government officials. There has been much talk about wicked curses states that will bring forth a mulatto politician. You probably remember some of these childhood folktales." Franklin turns around to look at some of the other employees in the cafeteria.

The people at the tables remain quiet for a moment. Then another employee says, "Yeah, were there were real devils sent south to destroy everything in Atlanta and other southern cities. But what is a demon or evil immortal? And are shamans real? Can they teach rituals to others and turn men and women into specters or zombies? Are they pushing other shamans or metaphysical people into all of people and us around? The way you can tell a person is possessed is if one eye, or sometimes both, has a distant stare. Remember, not all malevolent demons were made in Luxor and Thebes, in Egypt! No, some of the genies were designed in Meroë, Sudan. But priests have said that the Egyptians made too many metaphysical people. I've talked to many people and read books." After he finishes speaking, the employee starts chewing his food.

Dennis remains quiet as he listens to the other workers converse. At one point, Dennis begins to speak, saying, "Yeah, well, this talk about demons and witchcraft is pretty scary to me. Unbelievably, to this day, there are still whispers in Olney and around Greenwich about the demonic Minerva witch who walks along Interstate 87. I believe that's the specter who rides a shaman's broom through time. Yup, she may help us understand the Antichrist genies!

"You know the witch who haunts Minerva, New York? They whisper about how she used to grab the hand of an injured and bleeding captive on the Virginian plantations and lead them through Pennsylvania and into New York to safety. They say that this old demonic spirit possesses your body and that she hears and sees everyone, not just black people. They still say that she lingers around and haunts the small towns of Greenwich and Old Fort Niagara, New York. If you ask some of the Amish planters who have had family in those areas for some time, they will tell you that they ask, 'Who is whispering to me? Is it you, Minerva? Bring the young man who can see me!' Is the one that all politicians should know?" Dennis says to the cafeteria workers.

As Dennis is speaking, an elderly black woman gently strolls out of the kitchen area and takes a seat at one of the cafeteria tables. Josh joins her. Dennis realizes that this must be Marie, the Cheval witch. Most everybody speaks or tips their hats and whispers, "Hello, Marie."

Marie acknowledges the crowd by slightly tipping her head. She says to Dennis,

"Well, I might be able to help you with some of your questions. Josh told me that you may remember me from when you were a young kid. I know I remember your face. Did they foretell of a Ptolemy and of Obama? And what did Ulysses S. Grant tell my great-grandmother about Artemis and her twin? People who study black magic know that the succubus and the Medusa genies follow women who call themselves Minerva. Does Minerva and the 18 Devils support the 'Great Satan' hologram?

"Was there anyone calling themselves Minerva in New York around the time of September 11, 2001? It usually possesses women around the Poughkeepsie, New York, area. Many historical writers have mentioned a Minerva in New York. I do know that the elders have said that Minerva is an outspoken sorceress." Looking at Josh, Marie says, "But before I go any further, Josh, can you bring me a glass of water?"

Josh immediately jumps up and goes into the kitchen. That's when Marie straightens up and says:

Chapter 4

Highly Classified Watergate Testimonies From '72 – '74

The arrival of the Devilish Sorcerers and Demons from the ancient city of Hattusa, Turkey! {USA/Hatti or the USA government}.

1519-1915: The deadly Succubi of Babylon and the 33rd Legion move to the Americas?

{Confidential warnings from the Argentinian Dictator, Juan Peron, in 1972? The approach of Venus and the Azteca Quetzalcoatl god born in 947 or 974! Did the invasion of Constantinople in the 13th century lead to the discovery of Montezuma's Aztecas?}

"I hear you guys want to know about these Aztec 2012 - 21 curses or strive to become self-proclaimed paranormal researchers? First, you need to understand the Septuagint, a Greek form of the Hebrew bible. Professionals from different occupations and average thrill-seekers from around the globe should know that the Septuagint and other forms predate King

James Vi's versions. But I have worked with others before on supernatural activities and discoveries.

I was just talking to Mr. Franklin yesterday about the weird paranormal testimonials from the late Argentinian Dictator, Juan Peron 1895-1974. Back in 1974, many people throughout the Mexican Delta and Mesoamerica were worried about the full-scale war in Cuba, Mexico, Nicaragua, and Argentina. Some used the dangerous 33rd legion and believed the Aztec's Quetzalcoatl demons would arrive back to Central America between 1947 –'74. Certain people who worship the Greek Delta oracles still believed in demons and tried to prevent more dangerous military coups!

Many Mesoamericans claim to be a part of the 33rd legion and are convinced that Father Bernardino Sahagun should bring the Ptolemy's Italian and Greek family, which can communicate with oracles.

But Mr. Franklin claims that Juan Peron and others have proof the Mayan and Aztecs purposely prophesied about the deadly 20 Medusas and the strange paranormal activity of the 2012 curse. There were also many clues found that connected the discovery of the United States and the ancient magical city in Hattusa, Turkey.

Mr. Franklin and others nicknamed the city USA/Hatti or the cursed government of demons and devils. Some Mexican archaeologists support the theorem of Father Bernardino de Sahagun that Cortez and others were summoned to the Americas by Aztec priests! These historical claims may be supported by the Florentine Codex of 1570-'85. The Florentine Codex is somewhat of a 16th century Mexican/Hispanic history book of Mesoamerica. I've started to read volume 12 and investigate 'The Nahuatl Omen.' Some occult believes the reason that the Mexican and Spanish Inquisitions of 1570-1820 was fueled by these chapters! The Spanish bishop and governor of the Yucatan strongly knew that Maya, Aztecs, and even some Spanish believed in the 2012-21 curse!

Montezuma left word that his priest channeled someone within twenty years of 2012 and sixty-five years before! But they described it at the arrival of Venus, the Quetzalcoatl god, and their demons to Central Americas! I can clearly see a link between the ancient Mesopotamian governments which were later claimed by Greece and Italy.

But as you know, Watergate testimonies have proven that throughout the '60s and mid-'70s, Spanish, American, and European politicians were drugged and plagued with sex scandals by the American government! These politicians and towns leaders were hushed up when it came to the Aztec curse of 2012-21, the Nostradamus predictions, and Britain's 5/11 curse!

Franklin had information that *Really May Blow your Mind* when I researched President Nixon's Watergate testimonials from the nightly news in 1973- 1975. This was immediately after Lynne Fromme attempted assassination in 1975. A couple of news correspondents, along with Harry Reasoner, noted some weird affidavits about future tribunals. Wow, the name Obama clearly was stated in the news clip of 1975! A news correspondent was talking about 'Obama being president soon,' maybe this weird 2012 Mesoamerican curse! The news correspondent was telling the country that the Obama predictions and other psychic predictions were all simple little fantasies. Were these predictions scripted by 'Saturday Night Live' in the 1975-1976 President Ford and Carter debate? I still remember politician's testimonials". Marie says before, Mr. Franklin raises his hand and says:

"Yes, I clearly recall some of the Los Angles elite socialites and moviemakers who told President Nixon and others to destroy and hide all documents from the public! Some of these aristocrats from Europe and LA, California, wanted these unfortunate events to occur! And if anyone who'd disclose any information would be buffooned, poisoned, or murdered. Princess Grace of Monaco's visit to Philadelphia, PA, in 1976 should be studied for any proof of this dishonesty. Princess Grace and President Carter warned somebody in Philadelphia to seek help for Princess Diana of Wales and herself because they knew too much!

Yessir… Some of Princess Grace's Hollywood interviews from the '60s and '70s clearly state that she'd experienced bad nightmares accompanied by depression. Rumors circulated that in one week while battling her deep depression, she had similar weird reoccurring dreams of shouting warlike tribal Indians. By the weekend, she received a strange 'Native American tomahawk' in the mail. She openly warned the public that her dreams stated: 'You should not have any royalty unless the Native Americans are back! Did the Lakota Sioux Indians and others use Egypt's 'Insect People' in their religious ceremonies? Did Native American ritualistic ceremonies from the 1870s cause the Watergate controversy of 1974? Remember, the key to black magic is the demons created by Egypt's Insect people. Sounds peculiar, but I just googled a video of the 1974 UFO smokescreen on my cell phone.

Unbelievable respectable historians have written about these extraterrestrial legends. Although, there's a 2,000-year Ptolemy folklore that implies his family would reappear from the past with these UFOs or Anunnaki. Followers of Horus and Ra were to witness this sighting in a field outside Harrisburg, PA, in 2015! Reportedly, these Anunnaki, Beelzebub, and their demons have destroyed human civilizations every 2,400 -31,000 years. Sounds like a real fantasy, but there are writings from 1975 about 2012, '15, and 2018 UFO sightings for the Harrisburg, PA area.

Some celebrities in the 1970s evidently believed the notorious Montezuma and Tecumseh's curses. Actual warnings from Montezuma's and Tecumseh's tribal members, such as: 'We're taking everything back,' were frequently used in common street slang! I really wanted to investigate the late President James A. Garfield assassination on September 19, 1881. There were many rumors of a great world tragedy about sixty- six years later, or 1947!

Was Princess Grace of Monaco's car accident predicted to occur 15 years before 1997 or the twenty-first century? That would make it 15 years before the '97th year or a part of the King James 1597 curse! That would explain the Gowrie Plot on King James in 1600 and the weird satanic tattoo on the Ruthven body. But Mr. Franklin, Josh, and others would have to examine Princess Grace's testimonials. But there's also a very sinister prediction of a 2019 viral plague to match the 1519 Azteca smallpox plague along with a 2020 police brutality protest.

I've read Geoff Shepard's book and analyzed tapes on the 'Mysteries of Watergate, 1974'. It's inconceivable to believe the American CIA conspired to use mind-altering drugs, prostitution, disease on elected world officials! American CIA and FBI officials involved in the Watergate testimonies tried to 'Hush up' foreign officials about other mysterious events. That's when Argentinian Dictator Juan Peron's name appears on the list of informants. I tried several times to find the conversation on my cell phone. Maybe you can help.

Juan Peron recounted four-hundred-year-old historical facts from the Archdiocese of Cordoba, Argentina. Mr. Peron publicly stated the Roman Catholic Church in Seville, Spain, and Venice knew about the Americas since the 13th-century hundreds of years before Christopher Columbus! Some modern historians believe that ancient Aztec priests used Hydra's genies and other psychic abilities to provide the Italians and the Moors of North African navigational coordinates to find Central America. The dangerous 11th Jinns from Rome sent the Moors into Spain to create Al Andalus in 711ad, or 7/11. The Knights of the Fourth Crusade in 1205 gave the church and other Venetians this information on the way back from the destruction of Constantinople in Anatolia.

To understand the creation of the Americas, you must read about the 'Cursed city of Hattusa, Anatolia.' Ancient peoples were told that Anunnaki and their demons were about to destroy the 30,000-year reign of Egypt and create a new home around Mesopotamia.

The Hattusa legend states that these demons would destroy their own country between 1915-1947 and then return to Mesoamerica. That sounds identical to the Quetzalcoatl legend of 947 -974. And from 1974 to the twenty-fifth century, these demons would meet up with the other half of the 33rd legion in North America. Evil sorcerers with the Anunnaki and their demons had to overcome these demonic curses in another city or continent. This would mean that the former Hatti tribes would have to worship the succubi and her twenty medusas of Babylon! This would include the respelling of the name Hattusa to the USA/Hatti or the USA government.

These USA tribal members could only unite in the new city of Constantinople, Turkey. After they would subdue these Eurasian governments, a new headquarters or a joint continental capital would be established in the 'New Continent.' This sounds like a fairy tale, but Italians, Greeks, and Ethiopians forewarned people to not rebuild Hattusa, Turkey. Because the demons of the USA government will 'destroy all governments around them, even their own!' Just look at the destruction of Constantinople, Turkey, in 1205 and 1453. These succubi and their demons will create a 'Devils Apocalypse' every 3365 years. The Hatti of Turkey were written about in the Bible as the 'Caged Demonic Bird,' or flightless fowl!" Franklin says as Dennis clears his throat and begins to speak.

"Do you actually believe that the name Hatti will be caged in a weird Bermuda triangle in the New World and will remain oblivious to other nations? Dennis inquisitively asks,

Was the country of Haiti designed by ancient sorcerers to help the public study the occult? Makes me wonder about these dates, 1804-4018. But the Biblical quatrain of Daniel Chapter 2 and the falling metal Antichrist statue will reside among the succubi of Babylon, in the USA government!

Mr. Franklin believed that Daniel Chapter 4 would explain the fiery concentration camps in the 1940s and the events of Daniel Chapter 2 that occurred in 2001. Josh recounted that the Antichrist genies or the 20 Succubi of Babylon are at work. If you add the historical tragedies in Chapter 4 and Chapter 2, that equals 6. And both theological texts were supposed to occur 66 years apart, or 6+6+6.

Josh and Mr. Franklin both compared the former Argentinian Dictator Juan Peron's comments to the late Mrs. Denise Matthews or 'Vanity's' comment on demons. The late Mr. Juan Peron appeared to have this apprehension for these weird curses or demons. And I still remember Joan Rivers and Denise Matthews' discussion about demons who'd circle her. And about these are the same demons who'd be circling somebody within a couple of years of an approaching 2012 disaster. The late Dictator Juan Peron, Ms. Vanity, and Walter Cronkite all apparently believed that someone like Father Sahagun was believed to become 'President of the Unseen World' in 2012. This priest would explain Montezuma's II birth in 1466, and unnatural death in 1520. How was this 'Sammy' involved?" Dennis asks as the group remains silent for a couple of minutes. For about fifteen minutes, the group remained silent as they nibbled their food.

But before long, Franklin taps Josh on the hand and says:

"I still examine crucial events in 'Walter Cronkite 1972'. There are clear questions and talk about some type of near-future terrorist attack and a connection with the Twenty-first century. In the Kabbalah practice, the number 20 is connected to the deadly 20 Succubi genies of Babylon. Could Walter Cronkite be talking about the Twenty-first century and a 'President O' and his Omega genies who'd be pursuing the deadly 20 Succubi demons and the 13th angel? Does Greek or Aztec mythology explain the 969 sequences of jinns? Franklin asks while sipping his glass of water.

Dennis quietly nods his head while he dials cell phone and says:

"I've heard of the Aztec Indian curse of 2011 and the arrival of the Quetzalcoatl god, Venus, and their demons before. There have always been rumors of Montezuma II's revenge and the viruses of 1519 -1915 throughout Arizona, Texas, and Mexico. And I can remember a viral outbreak in 2019. Since 1861 and earlier during their Cinco de Mayo celebrations, they've whispered about Aztec black magic curses and demons. During their Cinco de Mayo celebrations or the battle of Puebla de Los Angeles, rumors have circulated that their Quetzalcoatl devils misled the 6,000 well-armed French artillery to a certain defeat to an underdog of 2,000 indigenous men or of mixed ancestry! Some of the indigenous Indians, like former President Benito Juarez, strongly rationalized some type of demonic black magic at work. They've always been whispering, 'We're taking everything Back from the Invaders!' In the sixteenth century, Father Bernardino de Sahagun warned the public about certain types of Nahuatl religious practices.

Father Bernardino had noted in his book the Florentine Codex that some of the Nahua religious practices appeared to be satanic in nature! Surprisingly, now some of these Aztecs' 'Human Sacrifices' for prosperity are now rising in the United States! Apparently, some believe that Mexico and North America a joined by the same demonic flag of the Quetzalcoatl demon, or the 'Eagle and the Snake'! Many South and Central Americans, not just Juan Peron, have discussed these weird and dangerous Aztec curses! I believe the actors, John Wayne and Lucille Ball, tried to create some kind of blended Native American family in 1976, who could hear these voices at night!

Franklin claims that many Mexican, Argentinians, and others have superstitious beliefs which pertain to the demons of the Quetzalcoatl god and even Zeus! But Constantinople and Egypt discovered the Americas! Some say that the Quetzalcoatl God and Venus are the same as the Succubi demons of Babylon. And these demons were supposed to arrive throughout the Americas within four hundred years after 1519. So, between 1519-1915, sets of Hattusa's demons would destroy their own governments before they came to North Americas in 1915- '47.

Unfortunately, the cursed city of Hattusa sits right in the middle of the old Armenian empire! What exactly came out of 1915 the Armenian genocide? But my question is what is planned for the dangerous year of 2020!' Dennis replies.

Dennis and Gloria converse briefly before Franklin answers:

My last visit through Turkey and parts of ancient Anatolia involved discussions on how these deathly wars began. In Turkey, they've always discussed their demonic leaders who'd demanded 'Human Sacrifices'! These sacrificial wars began with the 'Human Tails,' or people with dorsal cutaneous appendages. Supposedly, some occult followers believe that demons and gods dwell within the different types of human tails. There are pictures of the human tails with explanations of their mitochondrial genome which the Nazi's SS have used as literature on the internet.

But the Turkish Jews reported that certain German Third Reich party members were advised to move to Central and South Americas after the war between 1947 and '74. They were supposed to build a new Fatherland religion Argentina around these mystical black magic beliefs! There were many strange testimonials and confessions from former Nazi supporters, like Adolph Eichmann {Alias: Ricardo Klement} while being expedited back to Israel for trial. Many of them left testimonies of demons and the 'Arrival of Quetzalcoatl' back to Mesoamerica." Franklin says as Dennis intercedes.

Dennis immediately grabs his pen and says:

"These tales of 'Human Sacrifices' and worship of Baktun 13th are very similar to satanic cults which currently exist in Los Angeles, C.A. But some of the German people believe that Aztec priest put the '20 medusas' on us?' Juan Peron's statements appeared to spark in superstitious Mexican followers. The Mexican flag of the Crested Caracara with the snake in its mouth is remarkably a duplicate of the tiled Mosaic of an eagle with a snake engravened on the floor of the Imperial Palace of Constantinople, Turkey? Did ancient Aztec priests discuss this 2012 curse with Greeks from Constantinople, Turkey? Maybe, that was a bad omen? That's often the warning signal of a warlike regime transition of death into a rebirth. Well, some Byzantines believed that was the symbol of Christianity's control over the old gods or daemons. This has been a legend for the past twenty-five thousand years.

Remember, Mr. Franklin, when you stated that in certain Kabbalah practices, each century is presided over by different numerical demons? For example, you stated the 15th century was presided over by the '6 incubi Devils' and written as 1+5= 6 incubi. This would lead the destructive pathway for the 18th and 19th

Centuries governed by the 18 Devils! Did Bezaliel's 13th angel lead Constantinople to its destruction in the 13th century or 1204?

He also informed me that the Quetzalcoatl god was to be born in the year 974 or 9 +7+4=20, or a year before or after. The Venus god and her 20 medusas are to be born somewhere in the years of 1974, 1975-81. Allegedly, one of the deadly succubae was born six years after 1974. Mexica legend states that these demons were spawned around 1980-1981 or twenty years before the disastrous event of the twenty-first century. I believe some of these devils were born twenty years before 1974!

Were the Nostradamus predictions to occur in 2000-02 when the demons turn twenty or twenty-two? There're other deadly 20/20 predictions for the years of 2017 -21 when those demons turn forty or {20+20=40}. Ancient mystics described plagues and other weird murders?

In November of 1974 and in 1992, Darren McGavin and others apparently made some 2020 predictions of 'Get off my neck,' with an alleged case of police brutality. Mr. McGavin also discussed the Melvin Belli confession of 1969. Maybe, this is something the Black Life Matters social group should investigate?

You should forewarn others about the Centuries between 2001-2400 because those are when the Succubai of Babylon, Beelzebub, and the 18 devils become King! And the 33 legions of demons in North America and the 33 legions in Europe coalesce. I shudder to think about the District of Columbia's crossroads of rt. 66 and rt.95. Because within the next four hundred years they'll be even more deaths.

I want to hear more about the predictions and concerns you have about current political figures and celebrities... I could set up a televised phone call tomorrow with other historians around 10 am,' Dennis says.

The group converses for about an hour before they head upstairs to their bedrooms or continue to work in the cafeteria.

Chapter 5

Great Britain Investigates Weird Crimes

"Demons, Incubus, or Marids? Were Obama's People Called to Investigate Weird "Out of Your Body" Crimes!"

Death and Tragedy 15 and 18 years before 2012?

As Dennis, Gloria, and Sylvia head upstairs, they begin to discuss the history of the Onley, Virginia mansion. Sylvia says:

"Well, let me explain my historical family. Young children and young adults must understand that demons are a real threat to politics and governments. People should investigate legends of these Egyptian Insect people or Annunaki. I remember in 1966, elite citizens like Ossie Davis, and others from Natchez, Mississippi told me to introduce the young to these paranormal activities. This section was added to help the reader comprehend what Ptolemy's eye might see.

Nevertheless, around the close of the Civil War, some of Virginia's rich plantation owners got in contact with some of Grant's Union soldiers. There were weird rumors of the Antichrist genies and this weird reconstruction of 1866 or 966! Unusual rumors of a 24/56 apocalypse and destruction of everything sixty –five years after

1865 started to arise! William Cheval from France purchased these properties because they wanted to get his wife and kids away from the rumors of the 24/56 apocalypse by the Antichrist genies. There were rumors of a New Caesar in Britain with the destruction of Napoleon's government in France. The New Caesar should help slow Europe's destruction by the Antichrist genies. He'd served time in the French-controlled forts along the Ivory Coast, Africa, and was sent into Egypt to join Napoleon's army's conquests and pillaging.

"Once Napoleon was jailed, most of the French army was dismissed, leaving what remained in disarray. While being on military expeditions throughout the Barbary coast gave Mr. Cheval the opportunity to come to the famous land of Virginia from the Barbary Coast. And during his time on the Ivory Coast, Mr. Cheval begins to refer to Delmarva Delta as New Delphi, or Philadelpheus. People have always whispered about the Delta oracles with their temple to the North, which held their gods and genies.

"Mr. Cheval, having befriended his guides, invited their families to come and help with tending the land of Philadelpheus or New Delphi.

"The Chavals moved to Virginia shortly before Europe's civil war began. Rumors about the 18 devils or demons in Europe began to spread. People from the

Barbary coast had warned the population that the 18 Devils would leave Europe with war and famine. Death would befall anybody who tried to stop them!

"William Cheval produced two sons, William Cheval Jr. and Peter Cheval. Peter Cheval became a lawyer and moved away, which left William Jr. in control of the plantation once his father retired. Onley became a big shipping and industrial port along the eastern shore.

"Rise in immigration with the influx of different cultures and religions came a rise in the occult. Apparently, the Spanish bishop Diego de Landa was correct in his denunciation of the occult. Bishop de Landa was a sixteenth-century bishop of the Mayan Yucatan Delta. In Natchez, M.S., many plantation owners still read documents and articles concerning the Mexican Indian occult practices, the 2012 curse, and about the 'President of the Unseen World' in 2005! Many different peoples Native Americans and Hispanics occult followers brought their different practices of demons and gods.

Any type of witchcraft needs to be stopped immediately. People from the old country will tell you, "Never extend yourself!"

"'Rumors about Egypt's real demons are much too risky because they're too many of them. Apparitions like Seth, Pan, and Mercury are extraordinarily strong and will take over your body. You'll start to become a part of the dangerous 33rd legion when they 'see the Ptah' and have weird dreams. Then I'll know the demons have started to use your eyes. These demons have done that with a lot of people. And don't call on the beautiful white woman disguised as a little brown dancer.

"'These manifestations such as Seth and Mercury were made to be shrewd and very deceitful. Seth and Zeus are very strong that he can push over a onethousand-pound horse or temple pillars! The Jinn named Mercury is reportedly green in color and stands about six feet tall. He is very dishonest. The untrustworthy Mercury works with the demonic Romulus. These gods are reported to have possessed Napoleon and led his army to its destruction in Egypt. Some say that they can hop into your body and lead you to your death, screaming, "Now I'm going to make you know my ugly presence!"

"'My dad said that some of his comrades whom he had served with in Egypt have said that they won't kill, burn, or loot again because this kind of destructive behavior brings up demons that may haunt you. Some of these men still have nightmares of being burned alive and of things that pull at them ... and sometimes, something begins yelling at them while they're sleeping, saying, "I will lead you into Egypt and kill you! These Sphinxes are real, like all of us. The nose less

Sphinx is a sign of Rhinoculura's Demons. Now look at the Sphinx and remember 200 Watchers or the Antichrist genies have led you to your death! ... We can't help ourselves!" That's still the cry in some parts of Barbary Coast. Remember the demonic Sphinx, as it has always trembled the everyone. Egypt's Twenty-Seventh Assyrian Dynasty destroyed all old families {2+7=9}. The Ninth Angel, Bardiel, 'the humiliated son of god,' controls the number nine.

But all this discussion about the President of the Unseen World has been slightly fascinating. But I got get some sleep... I'll talk to you guys tomorrow," Sylvia says before she opens her bedroom door and goes to bed for the night.

As Dennis and Gloria shuffle to their room at the end of the hall, Dennis remembers important conversations from Harvey back at the Community College. So, as they enter the room and get prepared for bed, he says:

"Now, I recounted that Harvey said that the U.S. government is currently trying to obtain a priest or a psychic called the hierogrammat, which presently lives in the United States. This kind of priest trains them, writes their future, and instructs them to fly into humans or animals. But as you may guess, when you start mass killing anything, bad demons pop out!

You must study the Kabballah, 13th angel, and the -01 units to comprehend black magic. But that statue in Daniel Chapter 2 is a symbol of a demon called a Karabasan, or a Mare in English folklore. That story has been in my family for generations. The land was a route for the Underground Railroad in route to Poughkeepsie. Even today, many Native and Black Americans in Poughkeepsie, N.Y. still remember Artemis and other abolitionists. Ask about the usage of black magic on the Maryland and Virginia plantations.

"Now, I've heard other stories on plantations throughout the South. Theirs aren't so pleasant. Old European plantation owners said that they had strange killings on their lands. This sometimes involved black-on-black and white-on-white killing. In these cases, some of the culprits testified that spirits had hopped into and out of their bodies. Witnesses had said that the suspect became terribly angry and strong when this happened.

"Some plantations had people who complained that something talked to them at night and that strange voices had told them to spread viruses to others. They didn't understand the Dresden Codex so they'd sent for Father Bernardino de Sahagun's 'President of the Unseen World'! Remember Mexico was very important for the Northern embargo over the South.

Nevertheless, its rumored that this is what Mrs. Abe Lincoln complained about. Mrs. Lincoln reported that a certain psychic who could come out of his body threatened her. This psychic had said that a real demonic presence overtook his body and threatened to kill Mr. Lincoln in a séance. That same psychic had said that the force would also kill her kids with viruses. The psychic had also proclaimed that a married Navajo couple from an Arizona reservation would go North and blend in with the Maat Boy and the Cheval family around the area of Carlisle, PA. The Maat Boy could explain spirit gliding and see those demons that Osiris had made to protect the Navajo people. But there was a real warning about a woman who calls herself Minerva. 'Don't go into anybody because deadly demons follow and pop out of her!' It was all over the Maryland, Delaware, and Virginian plantations. That's why there's still an alert out for any woman calling herself Minerva, especially in the New York area. You may see some treacherous events follow her. That's what contributed to freedom in the U.S. It was much too scary for the owners to have slaves hopping out of their bodies and touching them."

"That is a lot of history to swallow. But what about this warning of giving viruses to the public in 2019. Did John Belushi give strange warnings of a virus? There were even predictions he would prematurely pass 19 years before 2001! I still research Saturday Night Live's 1976, 'weekend update, on the death of Chairman Mao … Room 219'. In 1976 New York detectives personally informed my family about the

2019 covid pandemic prediction! My grandfather had talked about the 2019 pandemic along with the great 1918 influenza epidemic after World War I!

But first, I'll grab a shower before I go to bed...." Dennis says to his girlfriend before he enters the bathroom.

Gloria, he keeps the conversation to a minimum and then goes to bed.

Chapter 6

Legend of The Barbary States: St. Veronica Of Caesarea Philippi

Prophecy of the 9th Crusade to the Holy Land? Saint Veronica will bring the New Caesar and Prince of Peace! Inscribed at the base of Djoser pyramid: Maat priests vs. Demons & Black Magic! Saint Michael, Lilith, and the Cult of Pan, in Banias, Israel… Hades' Prince of the Underworld'!

The next morning, Dennis is the last to awake. Sylvia is the first downstairs in the cafeteria for breakfast. Soon Gloria joins her. By the time Dennis is dressed and downstairs, it is eleven o'clock. As soon as Dennis walks into the cafeteria, Franklin and Josh wave and call him over to their table next to the kitchen door.

"Hey there, Dennis," Franklin says. "There's a breakfast table set for you. Gloria, Sylvia, and Marie are seated behind us. We have unearthed ancient artifacts from Palestine's fourth Crusade in the Holy Land in 1199 - 1204. History counts only nine crusades, but people view the current war in Afghanistan as another Holy War.

Religious monotheists have always discussed Daniel Chapter 2, the year 20042016, and the Iraqi Civil war. But many overlook the book of Daniel chapter 11: 36-45. Many kabbalah followers nickname this period the Salvation of the Antichrist governments. Historians have warned people of the dangerous three 9's or: {3+6= 9, the 9 verses in the middle, and 4+5= 9}. Governments marked with the dangerous 999/ and the 11 chapters of Nod are very dangerous!

They've always talked about a big war with the Antichrist and its 66 Legions within the Germanic tribes. Did a weird curse cause the 9/11 tragedy sixty-six years later? Was President Obama this the historic Othello or Maat priest wrote about at the bottom of the Djoser pyramid? Do these strange fables describe the same person? This man was to become president and beg for peace in the 'Home' of the Antichrist genies in the 21st or 24th century, or 2400/55!

Some followers of Daniel Chapter 2 and in Genesis 11:1-9 believe in the ancient book of Nod and Enoch. I've always noticed the similarities between the bible's Genesis and in the book of node's Chapter nine, verse eleven, or Exodus 9/11. Many have held conferences on the 11:9 threat of the antichrist's demons or the sighted ones living in the United States! Hydra and her demons are currently residing in the United States. Everyone must consider the documents of the twin towers with their metal statue in New Babylon. Is the United States of America the sinful New Babylon?

Some chronicle Palestine's Fourth Crusade as 11/18 because activists were trying to control the Antichrist Genies. Certain Kabbalah followers believed in the spiritual number nine. The bloodshed may end with the advent of the Ninth Crusade! But we need to focus on the occult beliefs of the destructive number eighteen. Is there something weird about the 18th year of any century?

Some Middle Eastern and Europeans share the same short parables of a young Jinn who becomes the New Caesar Philippi, a President Othello, and the Nostradamus prediction. Marie has many stories about the ancient city of Banias, Israel, the cult of Pan, and the two thousand Watcher genies in the twenty-first century. On some Middle Eastern occults, the year 2001 is written as 2401. They believe that A.D. started after the collapse of Thebes, Egypt in 653BC, or Memphis, in 655B.C., approximately 2400 years ago. Some believe that 2400/653 is when the Antichrist genies started their destruction, and that's the devil's apocalypse. We must immediately focus on the reversal of these numbers! And what's with the numbers: 1915 - 1519= 396, or 18. It's unimaginable to believe that the world will end in the year 3366 from natural catastrophes!

I believe Sir Isaac Newton tried to warn the Germanic King George I about Hitler and the approach of this 2001 apocalypse! I'm still trying to comprehend why some people read the Bible backwards by reading the book of Revelations by Daniel first. Are there 12 books after Daniel connected to the 11 deadly genies described in the Bible? There's clearly similarities of the 11, 12, and 13 genies from earlier Hellenistic philosophy and Christianity. Remember the 11 demonic genies should never cohabitate with the 13th genies because 11+13= 24 devils. That would explain why the book of Enoch and thirteenth angel was extracted from scriptures.

Do you believe we have about thirteen hundred and sixty-six years left if those numbers persist? Was the arrival of a 'Fair and equal judge', in 2012 explained in the Mayan Dresden and Grolier Codex's? Well, it's almost time for the midday shift change. We wanted to show you around the plantation and introduce you to the others before Josh, and I go home," Franklin says.

As Dennis walks over to Franklin and Josh, he notices that the cafeteria is crowded with people of different nationalities. He detects a blend of visitors and field workers. Once he reaches the table, he briefly shakes hands with both men, sits down, and waves to his girlfriend and mother.

Within seconds Franklin blurts out, "I hope you enjoy your breakfast. We were comparing the old rumors of captivity to Nostradamus's prediction while we were making everyone's breakfast. I'm eighty-six, and my great-grandfather worked these fields after getting off a slave ship from Sierra Leone.

"My great-grandfather was named 'Henry O.' The O stood for the place where he was captured, and that was an oracle school in Sierra Leone. Back in the 1840s, townspeople reported that Henry O used to frequently talk about the psychic schools or oracle universities that were on the coast of West Africa. I remember my father saying that Henry O had taught the kids how to use both halves of their brains. And if you exercise both halves of your brain, you can talk to the spirits while you sleep. He'd told them they had dolphin-like brains, as dolphins can swim and sleep at the same time. For example, the left side of the brain can sleep while the genie or spirit controls the right side of bodily functions. That's how we visualize the genie's origin. But most of the rumors about magic came from the major university back in Alexandria, Egypt, or Caesarea Philippi, Israel.

"Rumor back then had it that Henry O and others from the Ivory Coast noted that the people with vision tried to completely erase the memory of the hidden city of Banias, Israel. The genies destroyed the oracle schools. And tried to destroy any memory of the Ptolemys of Philadelphus and Caesarion. Back in those old Universities, people used to talk about Europe's future New Caesars in their future in their Northern palaces!

You must read about the Egyptian genies in the Book of Enoch. My father traveled to the ancient city of Banias, Israel, or Palestine. The city Caesarea Philippi was sometimes called Paneas of Palestine because of its worship of the cult of Pan. Some of their ancient monuments date all the way back to Alexander the Great! My father informed me that pilgrims of different Amen-Ra beliefs used to teach and pray at the Temple of Berenice Caesarea Philippi. The name of the temple means: Berenice's genies will bring the next Caesar, Philippi. Germanicus and his Roman legionaries have always written many stories about this time.

I've read some of the ancient Witchcraft and religious texts hidden in Siracusa, Sicily. Some of the old Italian books proclaim that Sorcerers told the Greeks like Antiochus I and III to build Northern cities into modern-day Turkey and Syria. They were to hide the Old World's demons. Some of the old documents had the beginning and end dates of old kingdoms. Just look at some of these ancient dates: [1] Jewish Khazaria of Kazakhstan would begin in the 6th century and end in 969A.D.. [2] The joint German and Russian governments would begin in 969 A.D. – 1866 A.D.= 897 years, {or 900 yrs.}. [3] If you are 66 years to 969, you'll get Hitler's government in 1935! Did the Nostradamus prediction originate from these ancient priests?[4] Remember, the year 1935 in certain occults is calculated as 2335, or 2+3+3+5= 13th angel. Is someone currently using Black Magic on the world's population?

Nevertheless, did Berenice's genies move the Holy Land north to avoid the 23/55 plagues on the Southlands. Some believe this is where 76 Griots used their black magic to form the '30/30 New World Governments' in Europe and the Americas two thousand years ago. The antichrist genies ended Cleopatra's thirtieth Dynasty. And their New Philadelphus government would resume on Route 30 in the New World! This new government should be clearly defined as 30+30=60.

Germanicus created certain genies among German religious groups to protect the older forms of worship and tradition around Lancaster County, Ohio.

"If you visit The Valley of the Temples, in Agrigento, Sicily, you'll hear old tales of the Genies of New Delphi. People describe how sorcerers-built cities in New Magna Graecia in the unknown continents of the Americas! I've visited Greece and much of the ancient sites noted in the 1957 movie 'Boy on a Dolphin' starring Sophia Loren and Alan Ladd. I've even met with some of Italy's movie directors. But I'm trying to recover any memories or legends of two young men sent to the United States, and one becomes president! Some believe that another young man was traced back to the Berenices of Greece.

Some Sicilians will tell you that quite a few people belonged to Sicily's ancient Cult of Janus. The cult of Janus used the Snake-Haired Gorgon, or the Medusas, to protect them. This cult predicted and instructed the building of the ancient city of Londinium or London, England, for Male protector god Ladon. They ensured the female goddesses or demons to protect Europe or New Magna Graecia in the Americas! Small remnants of that cult foretold rumors that the Graeae or the

Medusa's sister would use black magic to end Rome's control in the year 476 A.D. And that a New Magna Graecia would begin at the intersection of Route 476 and Route 30 in New Philadelphaeus, Ohio. Some Kabbalah cult members showed me how Route 476 and Route 30 intersect in Pennsylvania. And if traveling Northbound from that intersection, it'll take a right into Syracuse, in New York, on route 81, or {8+1=9}. Remember to Kabballah members the number 9 is meant to control the Antichrist Genies! Just read the lintel at the Temple of Seti I, in Abydos, Egypt: 'He who repulses the nine,' or the 1+8 Antichrist Genies being the enemies of Egypt! Unbelievable, Seti, his priest, and I tried to contain these 18 devilish genies in 1318 B.C.E.

The Janus Cult used Bezaliel's Thirteenth Watcher genies in black magic ceremonies to arrange Thirteen colonies in America. America's thirteen colonies were to begin 1300 years after the collapse of Rome in 476!

Egypt's old genies and demons had predicted the epilogue for the old Southern governments. Governments slowly decline as they approach the occult year of 24/56. Remember 240 divided by 18= 13. Because of the Thirteenth Angel, Europe's 24/56 prologue began in the 1100s with the Medieval Crusades.

Germanicus's 'New Caesar' would appear in London around the 1800s to help balance the untrustworthy genies. Within two thousand years, the Exclusive two hundred watcher genies would present of an 'Ethnic President Othello and a Prince of Peace in the New World. This period would occur after a thirty-six-year war with Satan's devils in the nineteen-forties. This epoch of time was noted as the 'Generation of Peace.'

I've always remembered my father saying: The Old Testament describes Egypt's the Old Kingdom. The New Testament describes Memphis, Egypt's New Kingdom, and Revelations describes the 2300/55-year reign of the Antichrist Genies.

In the ancient city of Paneas or Banis, Israel people converted the ancient genies and demons into Saints and Devils. Egyptians had sent word around the schools advising people to be careful which soothsayers they used because some had demons that couldn't control themselves and wanted to destroy everything.

"Alexander the Great had told Ptolemy's family to warn Soter-Berenice of demons who'd set it up so that his death would occur in Afghanistan. And those very same demons and sorceresses would be sending more people to deadly Afghanistan in the future. I believe in the late 1800s, Winston Churchill had heard something about the Soter-Berenice of the future. The English used to talk about him as he passed through Egypt on his way to fight in Afghanistan.

"Artemis's and Dionysus's would assemble the children in the future, who were across the sea in Atlantis. Ptolemy's eye would be accompanied by an ethnic President of the Unseen World. And Saint Michael and the Ptolemy would help comprehend historical soothsayers or psychics.

"Ancient aristocrats like Diogenes, Caesar, and Caesarion were believed to channel these children in the future. But demons were placed on the finders of the new lands. Germanicus and others had special demons made to control the unruly northern barbarians. Since the New Caesar, Germanicus moved north into Great Britain and Germany. Others were to be killed by Cyclops Brutus and his demons. But excessive knowledge of Black Magic is used by the cult of Pan in Banias, Israel, is dangerous.

"People were warned about the year 2400/56 after Memphis, Egypt. The Antichrist genies would bring great disaster throughout Sierra Leone and in the land of Philadelphus. Was slavery designed to end in 1866 or three years before 1869 or 969? In the Kabbalah occult, figure 1+8=9 is the sacred digits for 'The War's End'! King Mansa Musa had warned the soothsayers and the public.

"Descendants of ancient oracles had sent King Mansa Musa west to build for Queen Cleopatra V and Princess Berenice IV. As you know, these Egyptian corulers lived long before Henry O or Mansa Musa. Rumors had spread that they were to send Hercules's descendants and other children of the future to Atlantis. The future city of Philadelphus, Jamestown, VA., and Londinium beyond were built by Germanicus' 76 Griot Genies. But Cicero took control of Osiris and eventually destroyed the mysterious old system," Franklin says.

Dennis has been eating while listening to Franklin tell his story about his greatgrandfather Henry O. Now, Dennis is digesting his food along with this wild tale. He finds the story too mesmerizing to keep quiet. Before long, he decides to ask some questions. Dennis says, "Well, are there written facts or rumors about Saint Veronica bringing a new Caesar Philippi to Londinium to battle Hitler's Antichrist genies in the 194o's? And was St. Veronica Caesarea Philippi a real temple in Israel or Palestine? Some Palestinians have always acknowledged the presence of some religious Ptah. Some actually believe that even today, ancient Egyptians are still channeling this Berenice IV's relative." Dennis glances over at Josh, Marie, and Franklin.

Franklin gently nods his head and says, "I took a few classes of ancient history at Olney Community College. The priests of the past are all gone now, but there are still followers in Israel and Palestine, and some will translate their text. Some remember the creations of genies or demons out of fire atoms or photons. But the Ptah's are not Caesars, and control of the genies has always been a real controversy.

"In Certain areas of Banias, Israel, there's still a fascination with rebirth or reincarnation. The legend states that Phoenixville, Pennsylvania, was given the sign of the ancient Phoenix bird. You must follow Egypt Road, and thereon you shall find the hidden city we built through black magic. Some believe that this is where Cleopatra's temples are, but you can never be sure of the interpretation.

"Religious followers used to talk about the Ptolemy's connection to the future. Historians reported that aristocrats of the past tried to set up a stage performance for the children of the future. But the Blonde Medusas turned things tragic. Rumors of a 2018 royal marriage of a young prince and romances have always been the topic. When the children reached adulthood, they rewarded them with great riches and romance. But that bounty was short-lived, as it was taken away by the evil Medusa, or the little brown dancer, and her Osirian relative.

"Well, I don't know everything, but from 1948 until 1952, I was stationed at an army base in Alexandria, Egypt. And back then, a lot of Arabs, Italians, and Greeks had whispered about the movie *The Magic Carpet*, circa 1951, with Raymond Burr and John Agar. Back then some of their producers were easy to talk to. Because people were still talking about tales of mystical Minerva's magic flying machines and other things. The Romanov families had produced many films and paper works on future family events. And six years after World War II, there were whispers about some ethnic man and Berenice IV relative going to a distant land in the future! Now, I'm not sure of the location, but I do know that important film producers were asking around and collecting any information they could from everyone in the city. But there has always been talks about 'Black President' of the Unseen World, in the New World of Atlantis! I'll try to locate some of my old paperwork for you.

"Royal Italian families have advised people to investigate tales of unseen spirits hopping into and out of the bodies. In Banias, Israel, they've known about the 'Eye of Horus.' But did St. Veronica maneuver her genies to become president of the United States twenty-four thousand years after Memphis, Egypt, or 2400/55? Were the New Caesar supposed to control the Antichrist genies or the 18 devils from their Northern palaces? What a phenomenal fairy tale!" Franklin states.
Dennis quickly clears his throat and says, "Well, I've never been to Egypt or Italy and don't use Black Magic of any kind.

"Yes, that's right," Franklin resumes. "Sorcery and black magic. But what we call black magic was sophisticated science and physics! We need more groups to investigate this science." Franklin says.

Marie interjects and says, "Reading books about Gideon the genie with reason. They've always summoned those genies at Paneas, Israel! But some a very hard to control. Allegedly, Ptah's Eye dates all the way back to Imhotep and is a walking, talking Rosetta stone—a real historical treasure!" Marie says Franklin states, glancing at Josh and Marie.

The group stays in the cafeteria for about an hour more. Then Dennis, Josh, and Franklin leave to visit the historical sites. While Dennis takes notes as they stroll across the plantation, he's left to wonder, *are these spirits still hopping into people's bodies? Maybe that's what the people in Fishkill and Long Island are talking about now in 2016 when they say, "Please help. Something is still killing us!" Some of those old books may save lives. And are these the same phantoms that many presidents and celebrities whisper about? Was there a deadly 2019 prediction for a Jeffery Epstein? What kind of omnipresent force is lurking in Native Americans? Are they still running governments and making us kill each other like it's happening in San Bernardino, California; Connecticut; Cohocton, New York; and Maryland?*

I need to know if there's anybody else out there concerned about these strange murders and other incidents. If there is, then we really need other people's help to investigate these rumors. People shouldn't let these certain tragedies happen to their grandchildren. These demons are trying to kill everyone. People need to know. I must visit Banias, Israel, and Caesarea Philippi.

Dennis contemplates his options as he looks across the plantation and toward the horizon.

Chapter 7

Descendants Of Hispanics & Native Americans

What did Montezuma's Revenge of 2011 Foretell? Was London's Future Princesses Involved in Montezuma's Warnings? Did the Comanchero Rituals Use the "Chant of the Salvations" at Night? ... Prince Rogers Nelson and his Gemini Trio?

At about ten o'clock, Dennis, Josh, and Franklin walk back to the main house. Dennis has already phoned Gloria and told her to meet him back in their room. So, after he shakes hands with the guides, he quickly retreats to his room. On his way upstairs, he knows he must tell his girlfriend about the ancient Navajo and Comanche "Tribal Chant of the Salvations...."

So, as he approaches his bedroom door, Dennis already has his thoughts prepared. He taps on the door quietly and then enters and whispers:

"Yes, Gloria, I hope you're still awake! ... I just had some questions about the year 1999. I remembered that Theodore Roosevelt's occult followers had some fascination with this equation: '66 yrs. -1999 = 1933'. President Roosevelt and his staff put an alert out for this sighted-dybbuk who'd threatened him over Native American or Mexican bureaucracy. Did early Native Americans know anything about the dangerous 33rd legion of demons? And who did Moctezuma II's psychic channel in 2012?

Most Americans and Hispanics never heard of the Native American Alliance Meeting of November 6th, 1811, or 1863? And many Mexicans and Native Americans still recount Montezuma's pre tribunal visions and testimonials of 2012-2021. Some of Moctezuma's occult prophesy and warnings are very similar to Prince Rogers Nelson's 1982 album titled 1999 -2000! ...I've examined some of these ancient Constantinople texts and asked myself, 'Was Prince Rogers Nelson and Denise Matthews' 2016 unexpected departure predicted?' And what are the Gemini Trio of genies, and how do they operate in an average society?"

He finds Gloria in bed watching television. As he takes a seat on the bed, she turns down the TV and says:

"Yes. I personally know people who'd associated with Prince, the musician. Back in 1982, some people in Hollywood believed that the evil Baktun 13, Quetzalcoatl, and other Native American demons were involved in certain actors' and actresses' death! Do the dangerous 18 devils follow the 9th Jinni? If you scrutinize

Mr. Nelson's 1990 Tokyo tour, he was always talking about a virus being spread to others in 2016 -2019! This viral information also came from Daryl Gates, the Los Angeles police chief, in 1985.

But I remember 1811, or the Native American cursed '9/11' because it was the day before the battle of Tippecanoe in Indiana! Some historians call it the 6/11 of Native American culture. Some believe it was the beginning of the end of the Native American nomadic culture. Many Native Americans and Mexicans believe that evil spirits or demons were the real culprits of their cultural demise, and it started about two thousand years earlier! You must read the Dresden Codexes' and Montezuma's 1466 writings on 'Omens.' These writings stated that strange beings and dangerous entities from Italy and abroad were coming to the Americas in the aftermath of war in Europe! You have to understand Daniel's chapter 11 on the approach of the Antichrist in 1999 and the 20 medusas. I've tried to study the ancient Greek, Hebrew, and Aramaic texts, but many are still untranslated. Apparently, it is pertinent to read Bidirectional to study the Dextrosinistral writings of the Quran, Mayan 2012-21 curses, the legend of Othello, or President OBAMA. Was 11/9 written correctly and pronounced thousands of years ago?

I've read about Tecumseh's brother's strange November or September sixth prediction. Tenskwatawa of Prophetstown, Indiana, was believed to have predicted the town's demise in 1811- 1813. Some Kabballah followers label those years as the demon years of 18/11 or 18/13. Most reckon that's where the September 6th curse originates from.

But you told me over the phone that Franklin and Josh noted that in the Southwestern part of the United States, especially Arizona, there exists an 'Untold Part of History'! ... Franklin had explained that back in 1863, there was another Native American Tribal Assembly. Tribes like the Navajo, Hopi, Mexican, and many more were fighting for their lives against their demonic European attacker. In addition, you can't forget the Comanche's peyote rituals and their use of the 'Weeping Eye' genies. However, people have whispered that certain tribes like the Navajos allegedly used black magic to protect their citizens. Some of the tribes summoned their citizens, whom they called the 'Eternal Life People,' specters, or the 'Immortals.' These were the same demons or genies which destroyed the Azteca!" Gloria says as she looks at Dennis.

Dennis gently rubs his girlfriend's leg and says:

"Yes, Franklin mentioned that in today's society, there's Ptolemy's relative who can still hear them! Some Native Americans believe they hear their ancestors call for them at night, even in today's society! A few Navajo members still use their relatives 'Eye of Horoscopus' to foretell their future. His Navajo great-grandfather was a spiritual leader and named this young man Ptah, or the Prince of Peace!

"The older members of the Navajo nation always have grieved over the illegal activities that 'Evil Lucifer's Band of Demons' and the 'Grim Reaper' will be doing to the Ptah and his family .... Moreover, Princess Diana of Wales should've been warned about the depression those demons had given her, and Dale Harper, or Kanga! This is the same dybbuk that has possessed President Kennedy's wife or the suspicious warnings that surrounded Princess Grace of Monaco's car accident on 9/14/82.

"Did the Medusas make Princess Diana and the two New York musicians a precursor of the 9/11 tragedy? Something still pulled me to research her death in August 1997. Are Tenskwatawa's demons still destroying lives in the twenty-first century? Navajo leaders have said that Medusa's demons have always destroyed many English lives since King Henry I of England, in AD 1100. These demons have always led countries into civil war.

Some Navajo members had described the Nostradamus prediction as 'The Medusa's Return to Europe' and Asia! Native Americans and other Indigenous peoples of the US Southwest have always called it Montezuma's Revenge! People have said that demons from Europe will play with time and tragedy. They arranged for the beginning of WW2 in 1935 and 9/11/2001 to be sixty-five years apart. The destruction of the English kings began about 965 or 966 years ago, and that's the sign of the Antichrist genies!

"The old Spanish and English settlers have told Navajo spiritual leaders' stories about Europe's Antichrist genies' migration to the Americas. And they've reported that those demons have destroyed the Native American culture for the past 366 years. These very same demons have left a deathly streak in Europe for the last six hundred years, and those are the 18 Devils' numbers.

"But Franklin had stated that the German 'Dybbuk Box' reminded him of Karnak in Thebes, Egypt. Karnak's inner sanctum, or the room of Godliness, had a similar wooden box that held their demons and Gods. Some Brazilians who'd come before Columbus had warned Native Americans and everyone about future Northern disasters. Certain rumors in Central and South America had said that the 'People with Vision' or 'Genies' would set up a 'Disastrous Demonic Ceremony' in the future.

"The evil genies had set up a Karnak 'Deities and Demons Ceremony.' The ritual is performed at the arrival of a New Caesar! But who was Great Britain's New Caesar in the 1920s?

These ceremonies maintain that you must build two obelisks at the entrance of the city and unite the Ptah, or Amen, and their genies in their wooden box. The other steps include the alignment of the other genies or gods. During the ritual, the demons would begin to destroy the two obelisks, and Seth gives their deity the Amen, a disease.

"Some Brazilians had claimed that the evil genies' ceremonies would continue for about eighteen years. Within the eighteen years, the water genies would resurrect the Ptah, two buildings, and their practitioners. This sounds similar to New York's 9/11 events at the arrival of Germanicus's genies.

"But tribal members in the 1870s prophesied that there were to be other famous 'Children of the Future' that will fall victim to these 'demonic people'! ... Some of the Native Americans were warned back then about the usage of Black Magic and the 'writing of people's future'! ... The writing of people's future means that a witch doctor places demons or spreads evil around their victims as they mature. It means that they assure their prey doesn't stray. You must remember that demons have memory! ... This tragedy is believed to have occurred in Palestine.

"Even today, older members of the Navajo nation still remember his great grandparents' expedition to Carlisle, Pennsylvania, back in 1878. Carlisle police had reported that Ptolemy's great grandfather used to live outside in his backyard in a giant wigwam. And this was around the S. West Carlisle area.

"Their daughter became a 'showbiz dancer' and was supposed to bear a child with one of Osiris's descendants brought over from Alexandria, Egypt, or Greece. Out west, they even prepared a Native American wife for the future 'PtolemyBerenice'! Ulysses S. Grant and other generals of the Western Territories had waited upon his arrival out west! ... These government officials knew they had a big problem arising. They had to put a stop to these strange murders of citizens around the Indian reservations because Native Americans and Europeans were being killed by 'the possessed'! In addition, there were debauched predictions of future British Royalty. It's time the world investigates Princess Diana of Wales's death on 9/1997. Were there any similarities between the simulation car accident in the movie "Eye of the Devil?" That J. Lee Thompson movie was released 11/18/1966, or 'Prophetstown's 11/18'!

But this Carlisle, P.A., Navajo family's journey had been commemorated in a lot of the 1940–1960s Native American historical records in Hollywood, California ....Those tribes have always been told not to use Lucifer or his demons and don't make 'evil spirit gliders' again!

"But I'll let you get some sleep. We could talk about it in the morning ....Just let me know if anything touches or whispers to you. And if you hear 'The Salvation Chant' at night, please seek help! .... Now I'm ready for a hot shower before I get some rest. Good night," Dennis says as he kisses his girlfriend.

Chapter 8

Doomsday Prophecies?

The Science behind Metaphysics America in the year 336??

The week progresses slowly while Dennis and his family tour the plantation's grounds as he did when he was a kid. Dennis writes down everything Franklin, Marie, and Josh told him about this 9/11 presidential prophecy and this Ptah's arrival. By his fourth day, Dennis begins to have weird dreams about King Nebra's lands in the north and his water genies. Dennis knows that Native Americans named the state of Nebraska after its waterways, but what Indian folklore did European settlers repeat? Franklin told him that an Iowan Native American woman had folklore about King Nebra's water spirits or genies. It'd been rumored that the spirits told Native Americans to head north toward Iowa and Nebraska for shelter and protection.

Dennis is slightly restless as he awakens alone with weird thoughts of genies on his mind. Everybody has been previously informed of an early apple-picking contest today, so he knows Gloria should be in the orchards by now. Since he is the last one downstairs for breakfast, he quickly grabs a sandwich and heads for the orchards. The orchard begins twenty feet from the house and stretches about a half-mile away from the west side of the main house.

Franklin and Josh wave Dennis over to where they're bagging apples. Franklin instantly says, "Yes, we have Gloria here bagging and picking apples too. We were just telling her about other famous and infamous people who've had doomsday prophecies too. I still pray for the unfortunately misguided groups like the Branch Davidians of Waco, Texas. People have noted some of their leaders are believed to have had similar beliefs about a future Judgment Day. Some Davidian followers had reported that certain 'sighted' leaders had been receiving psychic messages from their gods about a 'great disaster' or imminent insurgence. If you check their videotapes, they've even tried to attempt media coverage and public statements about this rebellion. The disaster or revolution was predicted within a couple of years after the '93 incident in Waco. And there should be another young prophet coming to explain and help protect people. And some ex-members had recalled the word *Lucifer*, *demons*, or *imp* whispered around their compound."

"But were their leaders talking about the trinity of gods who move the Ptah?" Dennis asks, scratching his head. "Did they talk about dybbuks, the possessions by the Hidden Ones, who move around in bodies? You know what Theodore and other presidents had warned us about. And were there any water genies in

Egypt's Nile River? I'm not sure, but I'm beginning to have weird dreams. Maybe I should visit New Philadelphia or Wooster, Ohio, and ask about Ptolemy-Berenice."

Franklin immediately answers, "Well, Banias, Israel's worshipers may be trying to contact you from the past. Or could it be the area of Caesarea Philippi? Scholars have suggested that with the manipulation of the mystery system, a person can Transcend time and space, but he needs to be stressed about the dangers of channeling.

"But back in '93, I was on vacation, back in Sudan with my friend D'auda and two others. There are two disasters that I clearly remember. The 2/26/1993 World Trade Center truck bomb at 9:18 am. and the Branch Davidians in Waco, Texas. We watched the victims of Branch Davidians and the N.Y. truck bomb on television and wanted to help the children and adults.

Both incidents show signs of the presence of Marids. As D'auda watched the '93 Waco siege, he had concluded that some of the Davidians' leaders were male dybbuks or Jinni. The meaning of *dybbuks* translates into possession by a Jinni—or *genies* in our country. You need to check the meaning of *Jinni*, and you'll understand. D'auda believed that a big Marid, an incubus or Lucifer jinn, had possessed certain leaders of their group. Whenever you have the incubus Jinni, you have the evil Sila and succubus genies who follow.

"The term *Jinn* is mentioned in sura chapters of the Quran. This may sound crazy, but someone believed they still exist. Historically, in our part of the world, we call these possessions the Hidden Ones. Many Jinns were made for aristocrats, the wealthy, armies, and common people. Most people can't see Jinni but only observe and feel their presence. These group genies are used by psychics use to channel and summon.

"But there are strict rules and laws of the genies. The Jinns and genies mostly come out at night when humans sleep. Practitioners must do exactly what your Jinn says, or they have the right to kill you. Occult followers argue that humans made these metaphysical genies, and they didn't create humans. But these Jinns won't take orders from their human masters. Etymologists have noted that being nonhuman has led to the wars between man and demon. They will seek the young and vulnerable to summon through Ouija boards or channeling. Remember what their creators had said— 'I cannot control my creations any longer!'

Russian royalty throughout the early twentieth century issued warnings to pregnant women who used St. Petersburg psychics. They were urged not to allow psychics to leave their bodies and 'Go Into' them. And that psychic readings may result in miscarriages. And that certain psychics would destroy the male offspring of Russian and Prussian royalty. Someone should analyze the current Presidential administration's children. Are there any male children from the past three presidents?

"But these metaphysical creations still currently exist because it's difficult to destroy unseen matter. So, how can we control the big Marid Jinn in current world leaders? These genies have claimed wars with the Germans and designed the Mafia for the Italians.

"To become a 'sighted person,' you must summon a Vetala Jinn with exactly the correct number of subatomic electrons and protons. To understand genieism, you must know the technical names and charts for the Jinni and genies that were created.

"In the Branch Davidian compound, there were lots of children fathered by one male or single leader. This shows the presence of an incubus Jinn and its clan of imps. This type of Jinni is followed by smaller Jinni or genies like Lucifer, minor Marids, succubi, Silas, and many more." Franklin notices that Dennis wants to speak. So, he inquiringly gestures for Dennis to speak his mind.

Dennis scratches his head for a moment and says, "Yes, my father talked about this type of sexual misconduct in the S-Bahn murderer or serial killer in Nazi Germany. Many German citizens still talk about Paul Ogorzow today. He used to whisper about the strange voices in his head. He was a Nazi squad leader and a railway car worker. He'd hunt and stalk German women and randomly dispose of their bodies from moving trains. This behavior of sexual misconduct with the ability to congregate and lead a crowd is a sign of a genie cluster.

After being caught and under investigation, he'd admitted to having had spirits or demons within his body! German pathologists have deduced that the weird voices and the strange, pronounced body movements are due to demon activity. Two of his victims had claimed that the S-Bahn murderer called himself an 'Incubus Prototype.' Nazi hidden documents claim that the former squad leader talked to himself and believed in 'See Somebody' in America? And in 1941, there were closed testimonies of predictions of rape and murder Long Island, N.Y. area within the next sixty years or early 2000s!

German investigators appeared baffled about how much he and some of his victims knew about remote areas in Long Island, New York in America. Was the ex-squad leader a psychic with remote vision? Who and what did he see in America? There was a phrase that had come from this case: 'Someone should explain what he was.' Was Paul a part of the 200 watcher genies who'd joined the Nazis in the 18th year?' The German government even informed its Soviet counterparts about these predictions.

We need to create a plan to control these beings before they destroy our current governments. But many don't comprehend genieism. The Jinni called the major Marids are still upsetting governments."

Franklin responds, "Most Mediterranean people don't remember their ancient beliefs and practices of Jinni and genies. Priests wrote about humanlike Jinni in the Quran and in the Bible's Old Testament. And they wear the eternal life amulets, claim to be sighted but don't assimilate the meaning behind Amen, Amun, or the Ammann. What is an Amen-Ra follower, and do they believe in the genies?

The three men begin to wander down the apple orchard, the dialogue about the concentration camp and Russian videos Dennis brought with him.

Josh intervenes and says, "Talk about genies or Jinni sounds absurd, but most people appear to have Hidden Ones inside their bodies. Osama bin Laden shared some common genie behaviors. Both could assemble large crowds and people and appeared to have been 'sighted.' This ability of foresight and other characteristics shows the presence of the big Marid, Vetala, and those are signs of the incubus's clan of jinn.

"We need schools with Scientologist who understand how to utilize math and science. These universities can stop them before we reach the year 3366! Some occult followers believe 3366 is the year of a worldwide environmental apocalypse! Will the earth naturally destroy itself?

"I was reading some of the current news on genieism. These details have stated that Seth, Karabasan, and the incubus's clan of devils who used to dwell in the region of Eastern Europe have moved. Now, these evil Jinni and genies are currently reported to inhabit a family in the Los Angeles, California, area. They've made the 'City of Genies, or Demons' the protective haven. These Jinns are considered dangerous because they make males question their sexual preferences. They were created to destroy the captains or presidential families of governments. Allegedly, these Jinns were talked about on your 1974 DVDs, along with the dreadful prediction of a '94 high-profile murder case. The tapes mentioned a couple of current Los Angeles, California, models, and actresses—God rest and pray for their souls."

While the group wanders the perimeter of the apple orchard, Josh leads the conversation. Franklin listens intently as Dennis writes little notes.

"What's the American government and Hollywood keeping secret about these manifestations? Are NY musicians channeling the spirits like Odus and the two Genies of Alexandria?

"Are there malevolent unseen spirits doing weird things to our bodies. Were these the predictions of the 9/11 epilogue or the 24/55 prophecy?" Josh declares as the group quietly continues to stroll the orchard.

Chapter 9

The Dangerous Antichrist's Warning Of 186?!

The alignment of the Karabasans + Incubuses + Medusas.

Can the Ninth Jinn Control the Assassinations of the Wicked US Presidential Trio?

{4oth U.S. President Ronald Reagan's cautious statements to Walter Mondaleon10/7/84, Louisville, K.Y.}

Dennis and Gloria quickly retire to their bedroom after spending all day in the apple orchard. As Gloria sleeps, Dennis reviews all notes that Franklin and Josh have given him. While watching the 11:00 p.m. news, he receives a call from Franklin. Dennis answers his cell phone and says: "Well, Franklin, I'd expected you'd be sleeping now. Because after I finish recording my notes on my flash drive, I'm turning out the lights. Nevertheless, I was just rewriting some of my notes about Egypt's warning from the Valley of the Kings in Egypt! Do they still talk about the Se Osiris 'Sealed Letter, Berenice IV's relative and another presidential man amid the ancient artifacts of Cairo, and Alexandria, Egypt? ... This story sounds unbelievable, but a four-thousand-yearold warning about 'evil flying things' or genies—? The Se Osiris 'sealed letter' is predated before the Prophecies of Daniel and describes this snake or the Quetzalcoatl god!

"Some historians believe that the male and female Karabasans' alignment caused the act of slavery worldwide and the decimation of the ancient Italian aristocrats? That's why the Mexican Nahual priest insisted that the Italians, and Dresden, or Berlin, Germany receive a copy of the Codex in the 16th century! Did this devil trio of devils drag these 'Children of the Future' and their families to a distant land? I've read statements in the Codex's and in ancient writings on many Native Americans Indians would be caged by the 66th legions in the eye or the center of this wicked 'USA government'! ... I keep asking myself is this the state of Oklahoma?" Dennis asks.

Franklin retorts, "Yes, some believe Akhenaton's and his priests' warnings about their metaphysical creations and 6+6+6 theory! They've always said that the Incubuses and Karabasan genies have made their political figures do weird things. Eventually, they will make anyone destroy themselves and their government. We call it the alignment of 18 genies or devils.

"And if you're African, Italian, Greek, or Middle Eastern, you should remember 'Juneteenth Day of 1865'! Some Sudanese shudder from fear at the American Civil War end date of 4/1865. Were ancient sorcerers forced by their demons to design 'hellish slavery' to end four sacrificial months before 1866? Remember, in certain Texan towns, slavery is celebrated as 'Juneteenth Day.' That means it ended around six months until 1866, or 9+6+6! Hell, that could explain the

Reconstruction era! Because those numbers amount to the 18 Devilish Genies and Jinns who call themselves the 'Antichrist' {8+1} =9! Old Civil War generals always whispered that devilish manifestations forced them to end slavery. Nobody should laugh when you hear the gossip of the Genie with the 'Off-Set Eye' or their Jinns. "Do you recall Ronald Regan's and others conversations about the 'Curse of the Wicked Presidential Trio'? Some skeptics believe the modern U.S. government is 'spellbound or hexed' and forced to resemble the American presidential administrations of Abraham Lincoln, Andrew Johnson, William Seward, and the sacrificial disaster of the Civil War.

"The '9+6+6 Presidential Trio' curse states that the first or second president enduring the Sacrificial Disaster may be sacrificed. After the first presidency, the second president may be slightly wounded, and the third president may face the impeachment trials. Egyptian and Sudanese priests designed the Incubuses and the male and female Karabasans to destroy governments by creating Civil War. Unfortunately, the Medusas and other entities even destroyed their own government and its people. Ancient Egyptian history appears weird when read by people who don't know about the genies. Did the genies or people with

'Vision' cause the big Rhinocolura's prison outbreak? And what force made the whole country turn Greek within a couple of years?

"I've been trying to warn Head of state Donald Trump's administration about the curse. Because historical data shows, the impeachment trials were headed for the third president elected after the 9/11 sacrificial disaster. Some say the six evil genies and Jinns tried to assassinate his political campaign, as they did Andrew Johnson in 2/1868.

"I still reread history about wars in Iran and Iraq in the '80s and '90s. Some view President Reagan, George H.W. Bush, and Clinton as the other 'Wicked Trio.' President Reagan was the first in that trio after the Vietnam sacrificial disaster who endured the '81 assassination attempt. And Bill Clinton was the third who was hit with the impeachment trials. I'm sure you recollect some of the Clinton files?"

Franklin pauses while Dennis continues. "Ahh, yes, I'm interested in this 'Evil Trio of 6 Curse' on former or recent presidents. But can you link these manifestations to the Valley of the Kings in Egypt or Sudan? I've remembered you've said that a certain Incubus prototype was known as the Grim Reaper, or the 'Off-Set Eye' Genie, which is still thought to inhabit San Bernardino, California. The word Incubus and the nickname Grim Reaper were whispered in 1960s Sacramento, California.

I've known people who've played with Ouija boards and received weird messages."

Franklin quickly responds: "You shouldn't try to channel or negotiate with a 'Blinking-Eye' genie, but you can with the Grim Reaper Jinn. That 'Blinking-Eye' genie is dangerous and is currently thought to inhabit the LA, California, area. The Genie had told Akhenaton to 'Shut up .... I will bash your son in the back of the head and take his life! Time doesn't matter; all presidential families should remember me, the "Off-Set Eye"! Especially when I will take your son with me to Hell!' I still tremble when I hear about President Lincoln's and Kennedy's gunshot wounds to the back of the head.

"But Gerald Ford's administration had left warnings about that particular genie or imp with the 'Thawed-Eye.' These warnings were to people in Ptolemy-Berenice's time. In 1974–1975 they had a talk show, in Los Angeles, California, about PtolemyBerenice, demons, and Gerald Ford's attempted assassinations! This information was important after the John F. Kennedy, Lyndon B. Johnson, and Richard Nixon trio. But these warnings talked about real demons or 'Flying Things' in California! Are these the same demons from Tecumseh's Shawnee of Ohio or Indiana? Was the Evil blonde Genie or 'Eve' with her 18 Devils in California?

"Los Angeles had Walter Cronkite and their newspapers advising everyone that demons were attacking political figures. Especially after the two failed attempted assassinations in September of 1975. As President Ford's administration cautioned the public, they studied the 'Trio of 9+6+6' and the Grim Reaper Specter! What really occurred three unholy years after 1966? Had Gerald Ford's administration even predicted President Reagan's shooting within six years of Sacramento, California? Ahh, yes, the American government knows about these manifestations in California," Franklin says before Dennis interposes.

Dennis pulls the phone closer to his ear and says, "Yes, I remember something about Evil Jinns and Genies who'd refer to themselves as '9+6+6.' You'd said that even Ulysses Grant, William Seward, and Andrew Johnson were warned of these manifestations! These were the same demons that many Italians had said came from Pandora's Box.

"Ahh, now I remember ... you talked about President Lincoln's hidden assassination details of William Seward and Andrew Johnson. Quite a few southerners believed the Union Presidents were devils and were cursed for what they'd done. But a small number of Confederates had talked about weird curses on them by people who'd called themselves demons or the 66 Legions of the antichrists. These 'People with Vision,' sighted Dybbuks, or genies, had leaders who'd refer to themselves as Karabasans and the Incubuses. In some places like

North Carolina, people took comfort in the occult and genies. Witch cults had developed and held beliefs that their political leaders should be sacrificed, especially after some sort of 'sacrificial disaster' like civil war or any radicalism. Did those beliefs lead to the assassination plot against Lincoln, Seward, and Johnson?

"Do you believe the 'sighted dybbuks' or the 18 devilish genies created the 'sacrificial disaster'–like Civil War? Did they align and surround the first president Lincoln with the six deadly genies? If you apply the 1866 presidential mythology to 2019, it will correspond to the 9/11 tragedy and George W. Bush. The great sacrificial disaster and the first president of the 'Unholy Trio' .... The curse's second and third presidents resemble former President Barack Obama and Head of State Donald Trump's question of impeachment! But please let me know more, Franklin," Dennis says.

Franklin quickly responds and says: "D'auda from Sudan just called me! He had data on William Seward and Andrew Johnson's 'Warning of 1866'! D'auda alleged that psychics who'd counseled Secretary of State William Seward had listened to his perpetrator Powell's testimony. William Seward had said that as Powell slashed at his body with his knife, he had thought he'd whispered, 'I'm an Incubus Prototype.' And they had cautioned Andrew Johnson about his future impeachment.

"McKinley, Theodore Roosevelt, and William Howard Taft were alerted to the '1866 Trio' curse before Taft's impeachment trials. When President Theodore Roosevelt visited Egypt, Sudan, and Kenya, their oracles nicknamed him 'Theodosius the Marid.' As Sudan's oracles explained the curse to him, they blessed him with the goddess Tamera, the good Genie. Egyptian oracles had even foretold of his cousin's Unholy Trio of Herbert Hoover, FDR, and Harry Truman, as well as the trio of Wilson, Harding, and Coolidge. I wanted more data on Harding's strange heart attack before an impeachment investigation. But they had projected the Eisenhower administration being the time of the Phoenix.

"Researchers who follow these 18 genies or the 'Evil Trio Curse' believe a Jinn or priest called the Ninth may help balance the 'wicked trio.' Sounds weird, but are there manifestations, demons, or possessions based in Los Angeles and other counties in California? When I questioned Californian aristocrats on the 9/11 tragedy, President Obama and Trump, they told me 'It was a set-up all wrong.'

"Were ancient sorcerers overthrown by their genies and forced to design slavery and war for their followers? Did the genies design the 9/11 tragedy and the Wicked Trio curse to be understood and controlled by something they call 'the Ninth'? In the ancient witchcraft books, the number 9 has always been connected to Regulating 9+6+6 or the 18 Devils. Some canvassers believe that these 'Antichrist Manifestations' have destroyed civilizations in Vietnam, Korea, and the Philippines. Should we wait for these demons to further decimate the rest of Mesopotamia and the lower Sahara?" Franklin says.

40th U.S. President Ronald Reagan cautious warnings to Walter Mondale, in 10/7/84. Did Tecumseh's curse begin in Fort Wayne, Indiana on9/11/1812? ... Is the 21st Century the beginning or genesis of a 'New Baktun'?

Franklin pauses for a minute while Dennis drops his head and searches through some old photos of past presidents. Within a couple of seconds, Dennis comes across photos of the President Reagan and Walter Mondale presidential debate of 10/7/84. Dennis is reminded of President Nixon's and Ford's 1974 assassination predictions for 1981, and 2015-2019! Almost immediately, some current dates pops into Dennis's head as he grips the phone, and says:

"Franklin, do you believe any of these theories which surround Tecumseh curse? Why the weirdness of the Fort Wayne, I.N. attack on 9/11/1812. Was the attempted Ford assassination on 9/5/1975, and the Gregory Lee Leingang premediated assault on President Trump 9/6/2017 approximately 42 years later both predicted? Didn't President McKinley's 'sighted Dybbuk' warn the public about this Shawnee, or Mayan curse in 9/3/1902? Theodore Roosevelt was about 42 -44 when the McKinley shooting, and his car accident occurred. The Richard Pryor psychic predictions of 1977 may've occurred 24 years later."

Dennis asks.

This question apparently catches Franklin off-guard. But as Franklin gathers his thoughts, he replies:

"Well, many personal friends have said that former president Reagan has always stated to Walter Mondale and publicly that he didn't want these weird presidential shootings to continue especially in 1984! Maybe, President Reagan heard of these Native American legends especially after Mr. Hinckley Jr. attempted attack in 1981, and since he grew up near Prophetstown, Illinois. But we should investigate the demonic tale behind Mr. Hinckley Jr.'s birthplace of Ardmore, Oklahoma. Some believe that local Native American tribes in Oklahoma predicted President Reagan's 3/30/1981 assault years before! Many Native Americans have always talked the activities of the President of the Unseen world in 2012-21! The legends from this area in Oklahoma should be examined for any predictions about George Floyd's, the Maat Boy and the late basketball player Kobe Bryant's accident. Were there any 'Demons' involved in any of these cases?

"Is there a weird 'eighteen-year tragedy timeline' before the 9/11/01 disaster? Just subtract two or three years before or after 2001 and evaluate the circumstances:

- The Ronald Reagan attempted assassination of 1981: '83.
- 18 devilish years before 1983: Robert and John Kennedy's assassinations of 1963: Malcolm X in 1965, and Martin Luther King, '68.
- 18 years before 1965: WW2 and Europe's calamities of 1947: '45.
- 18 devilish years before 1947: The Great Depression of 1929 and negative events leading to WW2.
- 18 years before 1929: President Theodore Roosevelt's shooting and WW1 proceedings of 1911-'18.
- 18 devilish years before 1911: President Garfield's assassination on 9/9/1881 and McKinley's assassination.

"We must always remember the words of the last Russian Romanov, 18th DynastyTsar Family (7/17/1918): "Weird out-of-body spirits were trying to kill us! Please bring our relatives who can see these apparitions!"

- 18 devilish years before 1881: The events of 1863, the Civil War, and President Lincoln's assassination of 1865.
- 18 years between the American Revolution of 1776: and the French Revolution of 1794.
- 18 devilish years after 1794: The War of 1812 and the Native American Removal Acts Wars.

" Are these 18 Devils real? Surely, these folklores and mythologies need more studies!" Franklin pronounces.

The two debate the facts for hours more before Dennis turns off the lights and gets to sleep.

Chapter 10

Tracking the 20 Witches' 5/11 Curse in The Twenty-First Century?

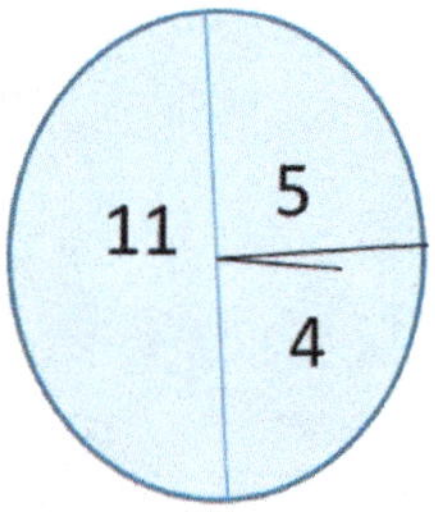

A mirror Image:

England's 5/11... {Guy Fawke's Day 11/4/1605} vs. New York Trade Center Truck Bombing? ... Or New York's 9/11, 2001 Victims of the New Homo Sapiens!

While Dennis, Josh, and Franklin stroll through the garden, they hand little notes to Dennis. Dennis places the collection of papers in his folder along with the rest of his earlier transcripts. As the group moves closer to the house, they begin to assemble on the front porch. While rewriting notes, Dennis receives a cell phone call from Bob, a close new coworker. Quickly answering his cell phone, he says:

"Hey, Bob, are you still rehabbing the old Tavern in East Anglia, Great Britain? The old term for that city was 'the land of the Angles.' However, did sorcerers rebuild the 'City of Genies' or Los Angeles in America for those witches or Scientologists?

I believe you've found some old amulets, Greek or Latin math books, and clothing from the 1603 period. Moreover, Matthew Hopkins and his witch-hunters bedded there during the medieval witch trials and the English Civil Wars. Nevertheless, tell me more about Great Britain's self-proclaimed 5/11. Was there a foiled plot to blow up Britain's Parliament in 1605 or Britain's 9/11?"

Dennis asks.

Bob quickly responds and says:

"Unbelievably, some of these townspeople have family roots from the medieval period of the witch hunt and witchcraft. Some of their families should put their history into a museum here. I found people whose families were involved in practicing sorcery and in witch-hunting. There were practitioners of other minor religions and the belief in out-of-body psychics. Many psychics were portrayed as witches or sighted-dybbuks. Some showed me necklaces and bracelets with hanging 9/11 amulets. The amulets had the numbers 11/5+4. Some celebrate Guy Fawkes Day on the fifth or the fourth. The British authorities arrested Guy Fawkes with the gunpowder on 11/4/1605.

"I've seen this geometrical figure when I studied algebra by Archimedes of Siracusa, Sicily. Villagers say the symbol means: 'The Gods or Demons encircle you on this date.' Nevertheless, how did poor villagers with limited education in the

1600s know about Archimedes? Maybe, legends from the crusades but some of these residents and their predecessors have never left the British Isles. Most of these Middle Eastern occult beliefs had come from Moors, the crusades, or Germanicus and his Legionaries. And other royal Roman family assimilations throughout England, Germany, and the Netherlands. Was that Shakespeare's inspiration for tales of Marc Antony and Cleopatra or the Berenices?

"The villagers told me stories about the Othello legend and their out-of-body psychics. Some believe the 9/11 terminology and Othello in 2005, or 2008 had come from them! Psychics on the British Isles have always talked about Shakespeare's Othello or some black man who would become 'President of the Unseen World'! Rumors from the 1600s had said that if you wear the amulet around the neck and in the mirror, the numbers 9/11 clearly appear.

"English and Irish villagers reported these rumors of the Omega genies: 'Nothing would stop the Second Disaster of 9/11, in New York!'—and that a President Othello, or President O, would become president three years after 1605, or four hundred years in the future. Followers of sorcery claim that the number 4 is the sign of the thirteenth angel and his two hundred followers! Some of these medieval predictions even include the testimony of war with the Italians and Hitler's demons!

"These sighted dybbuks or out-of-body psychics tried to declare sorcery as a justifiable religion! Some even called themselves demons and were burned and beaten. Most of them had said that they would put 'The Blonde Medusa' and her demons on the British government. And these demons will make other larger Amen-Ra religious groups retaliate for us! And many residents were warned about the fourth decade of any century, all the '40s. There are many tales that if you add the forties disaster to the British 5/11, you will have another Great Disaster in their sister country. Government officials didn't want Guy Fawkes Day to come to the Americas!

"Remember, the British government had orders to chase away those evil manifestations. So more than two hundred people were ensnared within eighteen years of one witch hunt. The sighted dybbuks swore vengeance on British royalty and its people. The dybbuks' curse was to begin about two hundred years after 1640 and every 18 years after. The local psychics may've predicted the French and Indian, American Revolution, and 1860s Civil War. Supposedly, those sighted dybbuks wanted vengeance for the witch hunter's subjugation of 100,000 or more people! And the deadly witch-hunting continued twenty years later and well into the 1700s. Even today, you can even find people who can testify to these threats.

"Many past Suffolk residents believed in the 1597 book Daemonologie, by Scotland's King James VI. But others had issues about the era of the Nostradamus prediction of 1555. Did the Dresden Codex influence Nostradamus predictions? Some English still believe that Nostradamus's great disaster occurred on 11/5/1605! But many believe that sighted-dybbuks and the Antichrist genies provoked Guy Fawkes's group into insurrection! Moreover, the antichrist genies had caused the 1605 Catholic rebellion against English Protestants.

"However, native Suffolk County people point to numerous minority religious groups throughout the 1400s to the 1600s. Scottish, English, and Irish villagers have told us to reread Shakespeare's Macbeth's 'three weird sisters.' The Earl of Rutland and King James VI had warned Shakespeare of plots to kill him and other political people during local witch hunts. Psychics had said that demons like the 20 Medusas and Eros and Poseidon were used in the 1597 threat at sea. Out-of-body psychics had told King James VI about the two 9/11 government plots and predictions during their many witch hunts.

"Still, neighbors had warned royalty about the ninety-seventh year. Read King James VI's warning to his Prince Charles I and their British relatives. King James VI believed that sighted dybbuks had cursed him! In addition, they would come out of their bodies and kill his wife in the ninety-seventh year! It was King James's psychics who'd left Great Britain with the three and four hundred warning of the ninety-seventh year! Was the death of an English princess a precursor of the 9/11 disaster? I rethink Princess Alice of Battenberg complained of an Incubus in 1930,

67 years before Princess Diana (9/6/1997). Did Neville Chamberlain inform Princess Diana of Wales about President McKinley's threat of September 6, 1901, and the black US musician and the ninety-six-year curse? Did these warnings scare the beloved Princess Diana of Wales? People should have alerted the British government at the funeral procession on September 6, 1997.

"Historical psychics had predicted Guy Fawkes's plot would fail, but nothing would stop the 'Great New York Tragedy' in the New World in 2001! The old psychics had based their formulas on cult activities and Macbeth's 'three weird evil sisters.' Psychics from the 1500s had forewarned King James VI that the Antichrist Genies birthed him eleven years after the 1555 Nostradamus prediction. King James VI was born in 1566 (or 666)! Medieval European and Mediterranean governments were alarmed at the creation of Great Britain by the Scottish and English merger. Was King James VI a servant of the Antichrist Genies? Does the 66th legions bring tragic curses every sixty-six years? Some Kabbalah followers believe that the 66 legions of demons have divided into 4 separate groups, or {66 divided by 4= 16.5}. This means that most members of the 66th legions will become leaders at the age of either 15, 16, or 61. To become a member of the 66th Legion at the age of 15 is very dangerous and should be watched!

Did the 666 curses explain the Great Plague of 1665, approximately 18 years after the English Civil War? And what happened to the eight lost European monarchies after WW2? Was Victoria's family scheduled to be the ninth monarchy?

"When I study the occult, I refer to the texts from Sonchus of Sais and the Book of Enoch from Egypt. Those books explain how to use the demons from Armenta and the Two Hundred Watchers. I've learned how someone used those triangles in America and in the Caribbean, such as the Bermuda Triangle! We've investigated many Shawnee members and Amish sects around New Philadelphia and Wooster, Ohio areas. This also included testimony from groups in the area of Memphis, Tennessee.

"Did the British prime minister Neville Chamberlain send a warning party over to the United States in the 1920s? Did he know about this deadly curse of the 'Thirteenth Angel'? The World Trade Center was built 31 years before its destruction in 2001. The number 31 reads 13 to occult practitioners who use mirrors to communicate with the 20 Medusas.

"Were there weird documents about East Anglia's demons giving viruses to the public in the German Ravensbrük concentration camp? In addition, some nickname Washington, DC, 'Satan's City' because the crossroads are route 66 and route 95. Some have joked that DC has the sign of the Devil's Crossroads of

6665 (or 6666 with one bad president).

Historians have found the Lambda Calculus symbol PY used throughout some of the old documents. Scientists reason that 00/PY is used because the number 0 begins every process in Lambda Calculus. Locals rationalize that these symbols and dates were used in local witchcraft rituals! The dates on the paperwork begin in 1199 -1204 with the Fourth Crusade to Jerusalem, Palestine. In some occult rituals, 1199 is reduced to a Lambda Calculus fraction over PY. So, the date 1199 is expressed as an equation of 18/11 or 6/11.

The dates are separated by four hundred years and seem to be a part of a series of futuristic tragedies! Just look at some of these predictions and dates: {.6/11=1204 ....5/11=1605 .... 9/11=2001 ....6/11=2401} ... or 6596. These witchcraft followers had written the first series of dates backward and used mirrors with the words Medusas inscribed in their ritualistic notes. And they've unearthed a second set of prophesized disastrous dates: 6/11=2801 .... 6/11=3201 ....9/11=3601 .... 5/11=4001, or 6695. Some occult followers believe these two series of 6696 add up to equal the antichrist's genies apocalypse of 2400/56!

Along with these series of numbers are historical maps of Jamestown, Virginia, a settlement by King James I on 12/6/1606. These maps clearly define heavily used trade routes from North, South, East, and West. Seemingly I've found occult maps with plans for settlements in Jamestown, VA in {12/6-6-6}, or Washington DC at rt.66 & rt.95 or [6695].

"Even today, English villagers marvel at the witch hunt predictions from 1618 to 1640. Some of these artifacts include threats of the World War I and II era. There are documents that declare a curse on famous witch hunter Matthew Hopkins. His opponents wanted to give him virus-like tuberculosis, or 'spit-up-filth'! The 'Unholy Trio Curse' is very apparent in the preceding reign of Charles I and the impeachment of Charles II of Britain. Unbelievable, there were sixty-six years from 2001 to 1935 and sixty-six years from 1935 to 1869, and that's 666! Didn't Egypt's 18th dynasty warn the public about these Antichrist Genies or the 66 Legion?

"Was there proof of a King James VI and the arrival of a President of the Unseen World? I'm still looking for proof of the 'Genies of Peace.' I'll send you documents about other smaller attacks on future communities later. Eventually, these sighted dybbuks will destroy everything! God bless the people of 5/2019, Sussex County, Virginia. The world needs the best demon hunters to help understand these threats before it's too late. What do you think of this algebra insignia 11/5+4?" Bob asks.

"I'll fax you the old documents they found. Some of them have the sign of the Amen-Ra cult or the double pyramid. Some occult uses these triangles to hold the

'House of the Armenta Fire Atoms.' They say those spirits follow these triangles! Later, I will show you how to use these pyramids for healing. But if you put two equilateral triangles together, and subtract the 2 unseen Watchers = 18 Devils," Bob implies.

16+4= 20 4-5=01 =2001

Dennis puts his cell phone on a conference call so Franklin and Josh can respond.

When trying to understand the occult, a student must always refer to King Djoser's Pyramid and the seven-year drought! Are the Antichrist Genies and the seven fallen Angels still causing drought, poverty, and civil war? Will the country still exist 1,366 years after 2000, or 3366? Is this the century of the '20 Medusas?' The little group converses about the year 4141 for an hour before they go in for lunch.

The Summation

I Miraculum Et Bendictionem

Is there a hidden message within the 66 books of the King James version of the Bible? And do the Chapters of Isaiah's Apocalypse and his Mark of the Beast have any real merit?

Was 1540 + 400 = 1940 considered a New Baktun? And what did Adolph Hitler tell his followers on November 8-9, 1923 in the Munich Beer Hall Putsch, or within his 24 years political career and dictatorship? Didn't the Dresden Codex forewarn Cortes of Europe's and Americas impending disasters on 11/8-9/1519? Did Juan Peron speak of some the arrival of some of Germany's Third Reich return to Central America, and Argentina?

Have the 1200-year-old legends of the Quetzalcoatl and the Xolotl gods come true? What's the real meaning behind the 9th, 11th, and 13 Lords of the Heavens, Hell, and Chaos? Will these demons return in the 2000's - 2100's and repeatedly war each other? Remember 2100-600=1500's.

We must investigate the 6th century Hindu Goddess 'Kali or Kaalifa', and her deities! Are these Oracles still operating in the E.Indies, Mexico, and California?

Did Los Angeles psychics forewarn Richard Pryor that he would expire shortly before the first Black President in 2005? And how did these psychics arrange the Future and why were they told to 'HUSH Things UP'? "

Made in the USA
Monee, IL
02 November 2026

67378095R00103